SUPERTALL

SUPERTALL

How the World's
Tallest Buildings
Are Reshaping Our
Cities and Our Lives

STEFAN AL

W. W. NORTON & COMPANY
Independent Publishers Since 1923

For information about permission to reproduce selections from this book, write to
Permissions, W. W. Norton & Company, Inc., 500 Fifth Avenue, New York, NY 10110

For information about special discounts for bulk purchases, please contact
W. W. Norton Special Sales at specialsales@wwnorton.com or 800-233-4830

Manufacturing by Lakeside Book Company

Book design by Daniel Lagin

Production manager: Julia Druskin

ISBN: 978-1-324-00641-1

W. W. Norton & Company, Inc., 500 Fifth Avenue, New York, N.Y. 10110

www.wwnorton.com

W. W. Norton & Company Ltd., 15 Carlisle Street, London W1D 3BS

1 2 3 4 5 6 7 8 9 0

For Maxine

CONTENTS

LIST OF ILLUSTRATIONS

SUPERTALL

INTRODUCTION

The Era of the Supertall

"Come, let us build ourselves a city, with a tower that
reaches to the heavens, so that we may make a name for
ourselves; otherwise we will be scattered over the face
of the whole earth."

Genesis 11:4

"A mouse by comparison," Frank Lloyd Wright noted in 1956
of the Empire State Building, at the time the world's tallest
building, in comparison to his own invention, a mile-high
skyscraper.[1] "I detest the boys fooling around and making their
buildings look like boxes. Why not design a building that really is
tall?" Wright's "cloudscraper," planned for Chicago, was going to be
the world's tallest structure, by far. It would be four times taller than
anything on the planet. "In it, will be consolidated all government
offices now scattered around Chicago."

Wright claimed that the building represented a milestone for
"the future of the tall building in the American city." If his plans were

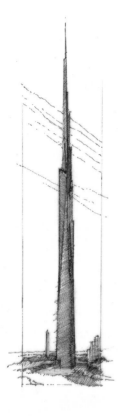

The Mile-High Illinois, Chicago,
Frank Lloyd Wright, 1956 (unbuilt)

applied to New York City, Manhattan could be razed "to one large green" with only a few mile-high buildings. You could "sweep New York into the Hudson and build two of them in Central Park and that would be the city." Ten buildings could "rehouse almost the [island's] entire office population."[2]

Wright raised both attention and eyebrows for his visionary proposal. He had no client, no site, and no budget. But the revered architect clearly had the credentials and the experience. At eighty-nine years old, he had designed hundreds of celebrated buildings. Was the Mile High a real possibility, or was it the hubris of the man who once identified himself in court as "the world's greatest living architect"?

As outrageous as the project may have seemed, it wasn't just a quick scribble. Wright held a press conference revealing a rendering that matched the scale of his vision, one rendering as tall as twenty-two feet, towering over the architect. He elaborated on the details of his massive "Sky-City," proposing landing pads for one hundred helicopters and parking spaces for 15,000 cars. He had distributed elevator banks throughout the schematic to ensure the timely evacuation of the building's 100,000 occupants within only one hour. He had specified that the 528-story building would be served by 76 yet-to-be-invented "atomic-powered" elevators, each capable of racing up to sixty miles per hour.

Despite Wright's immense reputation, no one dared to think of building the project. While the mile-high skyscraper did not come to fruition during Wright's days, we may see one in ours. In 1996, only four buildings were considered "supertall," classified by the Council on Tall Buildings and Urban Habitats as a tower exceeding 300 meters (984 feet). Today, there are more than one hundred seventy, with about a dozen completed every year—each taller than the Empire State Building. In Dubai, the Burj Khalifa, currently the tallest building on earth, stands twice the height of the Empire State Building, measuring more than half a mile high. The Jeddah Tower in Saudi Arabia is expected to be a full kilometer in height, about two-thirds of a mile. The new target may soon be the full mile high.

The sheer volume of Wright's building is now also within the realm of possibilities. Wright proposed 18,460,000 square feet of space, roughly 300 football fields crammed into one single tower. This is no longer record-breaking. In 2008, Dubai completed a new airport terminal roughly that size, complete with a hotel and mall, all under one roof, handling 85,000 people a day. Five years later, the city of Chengdu, in China, constructed a building that is bigger and perhaps even bolder: New Century Global Center is a shopping mall with offices, several hotels, an ice-skating rink, a Mediterranean village, and an artificial beach enlivened by a 500-foot-long screen displaying sunrises and sunsets.

What stopped us seventy years ago from erecting these megastructures, and why are we building them today? It wasn't the geometry of Wright's proposal. Surprisingly, both the Jeddah Tower and the Burj Khalifa resemble Wright's vision. The towers have a footprint shaped like a tripod, from which the building rises and tapers to a spire, giving it the silhouette of a minaret. Wright, who had enlisted the help of several leading engineers for his project, explained the virtue of this tripod shape. This triangular footprint is more stable, "the surest form of resistance" to the wind pressure, while the aerodynamic taper has its advantages as well. "It is really a steeple and has no wind pressure at the top." Wright's cloudscraper, as implausible as it may have seemed back then, made structural sense.

Wright himself thought his project was technologically feasible using the engineering advances of his time. After all, it was the time of the Space Race, the year the first satellite, Sputnik, was launched into space. However, Wright's project faced substantial technological obstacles. Even though he chose reinforced concrete, the same material of the Burj Khalifa, 1950s concrete could withstand compression of only 20 megapascals (MPa), about 2,900 pounds of pressure per square inch, enough to support a 20-story structure. While Wright sought to compensate for this in his design with a more stable, buttressed core structure, engineers doubted it would work. Today, breakthrough inventions make concrete stronger than ever. The addition of water-reducing polymers has increased concrete's compressive strength. This allows taller buildings with slimmer walls, as in the Burj Khalifa's 80 MPa concrete structure, a good 163 stories tall. The latest "ultra-high-performance" concrete mixes in steel fibers and exceeds 150 MPa, possibly reaching up to 250 MPa, about 26,000 pounds of pressure per square inch—the weight of three African elephants on an area the size of a postage stamp. Indeed, traditional cement has become quite the sophisticated blend.

But even with today's concrete, Wright's tower would have faced a constraint within the Mile High's very core—the *elevator* core. Circulating all the building's occupants with an average elevator of that era, traveling at the average speed of around 13 miles per hour, would require so many elevator banks that there would be virtually no room left for anything else. Wright sought to overcome the lift's limitations with nuclear power. His solution matched postwar optimism for an atomic future, like the Ford Nucleon concept car, which was envisioned to be powered by a rear-mounted small nuclear reactor, the same way as a nuclear submarine.

Today's elevators are still powered by electricity, but they are lighter, bigger, and move faster across thinner cables, up to 47 miles per hour. This outpaces the Empire State Building's initial 14 miles per hour by far. Elevator cabins can also be stacked to create double-deckers, doubling the capacity per elevator shaft. They also run more efficiently with traffic management algorithms that study usage patterns. This optimizes the grouping of passengers as they go to different floors, reducing the duration of each trip. The latest development is to run elevator cabins along magnetic rails, avoiding friction, like a magnetic levitation train. The humble elevator, once simply a steam-powered platform to hoist coal in mines, has become a modern-day marvel of engineering.

Even with all the technology of today's age, Wright's project would still face insurmountable hurdles. For instance, the 15,000 car-parking spaces would take up about 5 million square feet, the size of 100 football fields. Parking alone would consume almost a third of his entire building. Storing cars is not the most economical use of skyscrapers. Then there was the deep desire of postwar Americans, Wright included, for the *horizontal* city, not the vertical city. Americans escaped the dense downtowns for greener pastures: the suburbs, with their single-family homes and office parks. As successful an architect as Wright was, his Mile High building was a stretch too far.

The headwinds that buffeted Wright's "Sky-City" back then are blowing in the opposite direction today. What was once considered a castle in the sky is now becoming more commonplace.

IN 1919, DURING the Russian Civil War, Vladimir Lenin planned the world's tallest freestanding tower in Moscow with the aim to broadcast revolutionary messages far into the country's vast territory. The engineer he commissioned, Vladimir Grigoryevich Shukhov, was the perfect man for the job. Known as the Russian Edison for his numerous inventions, from grid-shell roofs to patents for oil refining, he had already pioneered lightweight tower structures. Prototypical towers of that time, like the Eiffel Tower, were built out of many parts, including heavy arches and trusses. By contrast, Shukhov's towers were built out of a latticework that formed a single, curvilinear surface, called hyperboloids. Imagine holding a batch of dry spaghetti from the top so that the ends fan out, creating a conical shape out of straight elements. When you fix all the pieces together with horizontal rings this structure becomes rigid. In his patent application, he had described the advantage. "The tower built in this way is a stable structure which resists extreme forces and uses very little material."[3] The ingenuity of the hyperboloids lay beyond their technical prowess. For their light, surface-like appearance, Shukhov called them "lace towers."

At 1,148 feet (350 meters), Shukhov's tower was not just going to be taller than *la dam de fer* (French for "the Iron Lady"), it was also going to be leaner. Instead of the many arches and trusses that make up the Eiffel Tower, which require many pieces of metal and joints, Shukhov's framework of diagonally intersecting metal required a lot less material. His structure would only require a steel order of roughly two thousand tons, whereas the heavy Eiffel weighed more than seven thousand tons.

But the Russian Civil War disrupted Shukhov's plan. "No Iron" read the entry of Shukhov's diary for September 20, 1920. "No Iron," he wrote again on October 1st. Construction accidents followed, including a cable that broke at the hoisting of a major tower section, damaging the already-built base of the tower. The regime called Shukhov in, warning him that the tower had to be completed or he would face the consequences. "Provisional verdict," Shukhov noted, "death by shooting."[4]

Shukhov eventually completed the tower, but at only half its planned size, at 525 feet (160 meters). He was celebrated as a Hero of Labor and awarded the Order of Lenin. The tower that would take his name, the Shukhov, did not live up to its original potential, however. And the hyperboloid, for that matter, followed a similar fate. While eventually more than two hundred hyperboloid structures would be commissioned across the Soviet Union, they were all smaller than the Shukhov. Outside the USSR, few architects dared to use this complicated geometry. When they did, they used it more as an accessory, like the architect Le Corbusier, who built one as a roof light on Delhi's parliament building. Instead of the marvels of modern architecture that Shukhov aspired for them to be, the structure that would become most associated with the hyperbolic shape became the nuclear reactor.

In 2004, I competed for the design commission of Guangzhou

TV Tower, at the time the world's tallest freestanding tower at 1,982 feet (604 meters). As a fresh architecture graduate, I had joined Information Based Architecture, a young design firm in Amsterdam founded by Dutch architects Barbara Kuit and Mark Hemel. We were bold enough to propose a hyperboloid structure. But our tower wasn't just going to be taller than the Shukhov. It was going to be more curvaceous. We deviated with elliptical sections, instead of circular ones. This gave the tower a more interesting profile, which looked different from every angle, with a skinny side and a wide side, in contrast to the Shukhov's uniform figure. We then twisted the top ellipse from the bottom one, which gave the tower a narrow waist. It made for a more recognizable icon—a shape that some people per-

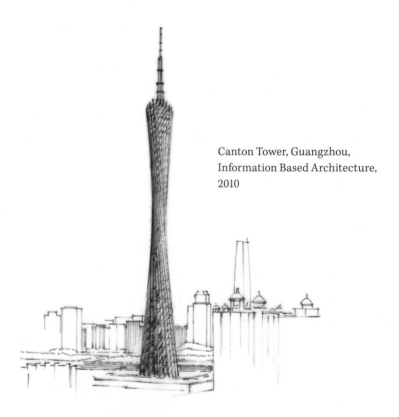

Canton Tower, Guangzhou,
Information Based Architecture,
2010

ceived to be distinctly feminine—and a welcome departure from the more phallic symbol type of large buildings.

Where Shukhov's tower was hollow, ours was filled to the brim with activities, including an observation deck, offices, an exhibition area, a food court, a revolving restaurant, and a cinema. There was even a spiraling open-air skywalk, allowing visitors to climb 590 feet up the waist of the tower. We also proposed a bubble tram around the top ellipse, spinning tourists in glass cabins.

To everyone's surprise, the Chinese clients picked our design from the competition. In 2010, our tower was built, then the world's tallest tower, a hyperboloid, with the world's highest observation deck. Why were we able to build this more complicated hyperboloid structure, at four times the height of the Shukhov, in an area of typhoon climate?

New technology made a difference. We were able to visualize the project's complex geometry with the latest digital design software and machine-aided manufacturing techniques such as 3D printers and laser cutters. Together with our engineers from the firm Arup, we built a computer model that was able to simulate loads and wind forces. This convinced our clients that our structure would hold up. Our model also helped test various iterations to reduce the number of joints, the most expensive part of the structure, making it more economical. We then built it with concrete-filled steel tubes, instead of Shukhov's bare steel beams, creating a stronger structure. We embedded hundreds of sensors into the tower, allegedly the world's first structural health-monitoring system, making sure it would continue to stand.

Technological advances alone do not explain the first hyperboloid megastructure. After all, who needs a TV-tower antenna in the age of cable and satellite? Almost as soon as our building was completed, our tower was no longer the world's only hyperboloid structure breaking the Shukhov barrier. In 2011, the hyperboloid Tower of Fortune was completed in Zhengzhou at 1,273 feet (388 meters). The following year, our tower was no longer the world's tallest free-standing tower either. The cylindrical Tokyo Skytree, at 2,080 feet (634 meters), while not a hyperboloid, was a good ten stories taller.

Even as Guangzhou's signature landmark, our brand-new tower diminished in prominence among its many rivals. A few years after completion, it was no longer the city's only tall structure, but one of a dozen 300-meter-plus buildings. As impressive as our tower may have been, with these and other towers quickly crowding the skyline, it became harder to stand out.

MOST CIVILIZATIONS HAVE an innate desire to reach toward the sky, to build the world's biggest buildings, like the ancient Egyptians built pyramids. Architects gladly oblige, following their utopian calling to build the perfect world, with a superior shape, in a flawless structure. But each society has its constraints. Wright's tower fought technological limitations and suburbanization. Shukhov's faced resource scarcity. Only rarely do all forces align that enable a society to construct unprecedented buildings. When they do, we call it a golden age, like when the Medici built Renaissance palaces in Florence, Italy. Or when, in the thirteenth century, Italian families competed to build the tallest home in San Gimignano. This eventually led to an entire town of tower-houses, some up to 200 feet tall.

Now we are witnessing another golden age, but not of *palazzos* or *casa torres*. This time around, we are living in the era of the "supertalls." In 2017, fifteen new buildings blew through the 300-meter supertall barrier. The following year, 2018, witnessed seventeen supertall buildings, including in cities that previously had none for good reasons. Consider the Salesforce Tower in San Francisco, on

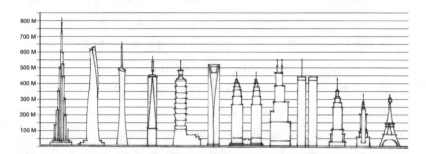

The world's tallest towers (selection), 1889–present.
Left to right: Burj Khalifa (2010), Shanghai Tower (2015), Canton Tower (2010),
One WTC (2014), Taipei 101 (2004), Shanghai World Financial Center (2008),
Petronas Towers (1999), Willis Tower (1973), World Trade Center (1973),
Empire State Building (1930), Chrysler Building (1930), Eiffel Tower (1889).

the lip of the San Andreas Fault line. In 2019 there were twenty-six
new supertall buildings, one of them in St. Petersburg, a city known
for its low, world-heritage skyline. The Lakhta Center, at the time
Europe's tallest tower, had been relocated seven miles from its initial central placement, only after UNESCO criticized the project for
looming over the historically protected skyline. Today, with more
than one hundred supertalls in the works from Melbourne to Moscow, there seems to be no stopping the supertall frenzy. Developers
around the world are locked in a supertall arms race.

Supertall is the story of today's tallest buildings. It traces their
existence to a convergence of two trends: the acceleration of technological progress and new societal preferences. The first is the
most obvious. Technology has always been an important factor
allowing for very large buildings. It was the invention of the safety
elevator that led to the skyscraper. Then structural steel enabled
builders to stack more floors on top of one another, upon which air-
conditioning allowed the ventilation of these higher floors. These

days, sensors monitor occupants and the environment to distribute conditioned air and electric light more efficiently. Machine-aided manufacturing enables designers to bypass blueprints and communicate directly with machines, allowing higher and more complex forms, such as curvilinear shapes optimized to reduce wind vortexes that the supertalls create. But the devil is in the details. At times, something as banal as window cleaning shapes how supertalls are made, to ease the herculean effort of cleaning thousands of windows at dizzying heights. Even the design of janitorial closets invites scrutiny, since every square inch counts when floor plans can get replicated more than a hundred times.

If technology is the seed, then today's "urban renaissance" is the fertile soil that grows the supertall. The twenty-first century is the first urban century. Never before has more than half of the world's population lived in cities, attracted by the many and diverse opportunities of urban life. When the influx of city residents exceeds the growth of urban areas, the law of supply and demand predicts higher land prices. Here the benefits of the tall building kick in. Since it maximizes the number of people on a given plot of land, it can offset the higher land cost. Meanwhile, it makes for more compact and dense urban living, allowing for more possible social connections between people, which can stimulate economic growth. But, as COVID-19 showed, there are downsides to density. Skyscraper living can be more contagious, with aerosols easily spread in crowded elevators and improperly ventilated high-rise shafts. The pandemic-induced urban exodus may grind the supertall outburst to a halt.

Some countries have decided to base their national policy on the positive relationship between urbanization and economic growth. China has added roughly half a billion people to its cities, and has the largest amount of supertall buildings. Some governments go so far as to urbanize deserts, like Dubai, its growth driven not by oil but by real estate. They bank on the risky mantra of "build it and they will come," as the Burj Khalifa showed. Hundreds of residential apartments sold out within hours. Skyscrapers are even altering historic cities previously inhospitable to urban growth. For three centuries, the skyline of London was defined by St Paul's Cathedral, with its silhouette of a dome and spires. Then, in the 2000s and 2010s, more unusual and often taller skyscrapers came in, crowding the skyline with ingenious shapes that earned them their nicknames, including the Gherkin, the Walkie-Talkie, and the Cheesegrater.

These two trends are accelerating. Automation, "smart" devices, and artificial intelligence are the new technological frontiers. These help achieve faster construction times, bigger operational efficiencies, and easier maintenance of massive buildings. The world's urban populations are expected to grow from 56 percent today to 68 percent in 2050, particularly due to growing cities in developing nations.[5] All of this could lead to even taller buildings.

New technology and unprecedented urbanization have made our society fundamentally different since Wright proposed his Mile High tower. What were once considered impossible ideas have become architectural opportunities. Today, it may just be a matter of time until one building goes the full mile.

THIS BOOK COVERS tall buildings and above all the *supertall* (above 300 meters or 984 feet), plus, in a few cases, the more loosely defined *super-large*, a multipurpose building containing most of the functions of a city, everything bigger than five million square feet (the size of one hundred football fields). When a building reaches this scale, it radically confounds the distinction between building and city. Imagine a world where streets, plazas, blocks, and entire buildings are absorbed into a single structure. For some people, this may be terrifying, maybe even leading them to cling to the ground. From an urbanistic perspective, however, it is unprecedented. More than just a building, this is a vertical city, where tens of thousands of people can live, work, and play in one single complex, without ever having to leave. The world of megastructures is fundamentally different from the world of regular buildings.

Decades before megastructures became a reality, avant-garde architects like Frank Lloyd Wright obsessed over structures this size. It wasn't all about height, however. Some toiled over highly efficient circulation, like the Italian futurist Antonio Sant'Elia, who in 1914 envisioned La Città Nuova ("The New City") as an array of monolithic skyscrapers articulated with snaking lifts, railways, and aerial bridges. Others expressed colonial fantasies, like the architect Le Corbusier. In 1933 he imagined a giant viaduct structure for French Algiers with fourteen floors of residential cells underneath the road for 180,000 colonized workers. There is the occasional playful structure, like the 1960s Plug-In-City, spawned by Sir Peter Cook and his fellow "Archigrammers." Here your living-capsule would be on the permanent move in a giant network of cranes. Some responded to environmental concerns. Paolo Soleri drew up eco-friendly buildings such as the 1969 Hexahedron City, a two-kilometer-tall beehive housing 170,000 people. But since none of these grandiose visions had been built, renowned architecture critic Reyner Banham, in his 1976 book *Megastructure*, declared the concept dead. He cautioned that it was probably for the better.

When I first visited Hong Kong, I was amazed by its skyline jam-packed with more skyscrapers than any other city in the world. They are lit up by lasers at night, and interconnected in a three-dimensional maze of skybridges and zigzagging escalators. All of it had been built without much hurrah, and certainly without manifesto. I saw places such as 26-story malls and live-work-play complexes built on train stations accommodating tens of thousands of people. I ended up briefly living in one.

Having served as a professor at various universities, I realized that none of the textbooks prepared students to fathom the sheer scale, density, and complexity of the reality of Hong Kong urbanism. When I moved to New York City, it was the same story all over again. I landed right in the middle of a new skyscraper boom with a dozen supertalls going up, a race for the highest observation deck, and a new breed of skyscrapers, so-called super slenders. Several of the city's tallest towers were designed by KPF, the company behind the design of four of the top ten tallest skyscrapers. I joined the firm for a few years. I found myself working on projects so big, they went beyond even the wildest dreams of the 1960s architects. Truth is stranger than fiction: that's the story of architecture today.

"We shape our buildings and thereafter they shape us," Winston Churchill once said. As we reach toward the mile height, what does all of this mean for us? As designers are drafting these vertical cities within cities as new additions to the skyline, they are also engineering new ways of living, working, and recreating—perhaps destined to be adopted in the design of other buildings. They are asking themselves the difficult question of how to design spaces for social interaction, overcoming the typical skyscraper shortcoming of segregating people in stacked floors. They need to create architecture that delights, despite myriad functional, mechanical, and structural constraints. They need to design the tallest structures while also humanizing them, making sure they don't overwhelm the streets below. Most importantly, they need to avoid the trappings of large

Robert Moses–scale projects, which were once so idealized but in hindsight wreaked havoc on historical cities and social communities.

As we are reaching the mile, are we stretching the limits? In New York, the city's tallest new towers are homes for the ultrarich, rising while the city's housing affordability has taken a plunge. At night, many of these apartments are dark, with their owners living in them only a few weeks a year, while every day the towers cast long shadows on the neighboring streets and parks. The truckloads of cement and tons of steel to build a few homes so high in the sky place a disproportionate amount of strain on our environments. COVID-19 threatened high-rises and dense urban streets, as hotbeds for the virus to spread. As ingenious as these structures may be, they are also markers of increased inequity and societal risk.

In this book, I tell the story of the world's tallest buildings. I will show the behind-the-scenes design and engineering making possible the monoliths we take for granted, from the 700-ton swinging weights which "tune" buildings to the double-decker elevators moving up to 50 miles per hour and without "howling." I will discuss the history of city planning in relation to skyscrapers, which exist on the intersection between fickle human wants and social constraints. I will uncover new futures for our taller buildings, including skyscrapers made of wood plus buildings that are entirely tree-covered. As I examine the resources, attitudes, and desires that have shaped how we imagine and build skyscrapers, I will ask the question: why should we build tall today, and is there a way to do it sustainably?

The supertall skyscrapers of today are triumphs of engineering, test beds of technology, and living laboratories of new ways of living. But in their efforts to break ceilings, they are also structures built on serious risk. One simple change of a window mullion alone can make the cost of a building skyrocket, sending the designers back to the drawing board. Or worse, extending a floor by a few feet can topple the dead load of a skyscraper to such an extent that it could lead to collapse.

We are living in an urban age where the most tangible architectural expression of this era is the supertall. They are the cathedrals of our time. The supertall represents a highly calculated, high-tech world, where not a square inch is left to chance. It is one that is highly interiorized and air-conditioned, and also frequently subservient to the rules and goals of the landlord, inhibiting activities, such as protesting, that might be possible in a traditional public space. It is a high-end world, a capitalist Who's Who of the most expensive and luxurious real estate available. But unlike the cathedrals of the Middle Ages, many of these new pinnacles in the skyline are off-limits to the vast majority. What would more egalitarian skylines look like?

The world of the supertall is at once a victory of mankind soaring to new heights and a controversial marker of society's inequity and climate change. It is a barometer of our society's highs and lows and a window into our future that may allow us to rethink our ways.

PART I

TECHNOLOGY

ONE

The Building Block That
Binds the World: Concrete

B urj Khalifa. This gravity-defying monolith in Dubai is so tall, it exceeds everything humans have built on this planet. On a clear day, you can see the tip of the spire from up to 60 miles away, before anything else in the city. The Burj (which is Arabic for "tower") is twice as high as the Empire State Building. Its sheer scale even transcends the earth's local temperature and climate. The exterior temperature at the top is thought to be 11 degrees Fahrenheit (6°C) cooler than at the base. The tower occasionally pierces clouds, with the top bathing in sunshine and the bottom suffering rain.

Time warps in the presence of such a vast structure. From the top, the view into the Arabian Desert reaches so far that the sun sets several minutes after the sunset viewed from the ground. Dubai clerics decided that the residents above the 80th floor should wait an additional two minutes to end their Ramadan fasting. And those above the 150th should wait an additional three.

None of this would have been possible without something we take for granted. At the core of the structure lies an invention that is underappreciated, maybe because it is ubiquitous or maybe because

Burj Khalifa, Dubai, SOM,
2010

it has been part of our lives for 10,000 years. Concrete. We use around 10 billion tons of concrete per year.[1] That's more than a ton of concrete per person, about twenty bathtubs full for each person on this planet. We have so much of it that if our civilization were suddenly to collapse, future archaeologists would find a thick layer of orange-colored rock on the earth's crust, its utter volume a testament to our concrete addiction, its shade a function of rusted steel.

The origins of the building block that binds the modern world can be dated back to the Stone Age.

Without concrete, the modern world would not function in the same way. There would be no Roman viaducts to carry water where it was needed, no dams to create reservoirs, no perfectly straight roads to allow fuel-efficient transportation, no floors to prevent hookworm larvae from burrowing into people's feet. There would be no Pantheon, no Hoover Dam, no Sydney Opera House. Even airplanes would not be as large, since runways would not be strong enough to withstand the repeated heavy traffic. There would still be skyscrapers, probably constructed out of steel, but they would not come close to the Burj's near-kilometer height.

A world without concrete may have been more sustainable. Concrete contains cement, a term that is often used interchangeably with concrete. Cement refers to concrete's essential component, the substance that binds it all together, like lime. Producing cement requires burning fossil fuel. The chemical process that creates cement emits carbon as well. Because of cement, concrete is culpable for 5–8 percent of the world's carbon emissions. This is twice as much as the entire airline industry.

This is the price we pay to modernize the world, and to help skyscrapers defy gravity. The higher we build, the bigger the gravitational pressure imparted from the higher floors onto the lower ones. This principle has long dictated the shape of our buildings. It led ancient architects to favor pyramids with wide foundations that supported lighter upper levels, like the pyramids of Giza. But such a solution does not translate to a city skyline, nor would it be efficient—a building a kilometer tall would be roughly one-and-a-half kilometers wide. Try to squeeze that into a city center.

To understand what makes supertalls such as the Burj Khalifa possible, and what gives them a relatively skinny silhouette, we need to appreciate the exceptionally strong concrete that was poured into its walls and foundation. The concrete in the Burj Khalifa can withstand about 8,000 tons of pressure per square meter—the weight of the Eiffel Tower on an area the size of a table. The future of concrete may lead us to a path of even higher strengths. Plus hopefully a more sustainably built world. And with new ways of manufacturing, including 3D printing, we may create even taller and more sophisticated buildings in the future. What horizons can be viewed from those future heights?

ARCHAEOLOGISTS BELIEVE A LIGHTNING BOLT led to the invention of lime, the essential binding substance of concrete. How else would our Paleolithic ancestors have discovered that heating up limestone to furnace-like temperatures would create a magical powdery substance? A campfire could not create a temperature that hot. But when our predecessors found out about lime's chemical properties, they must have been quickly impressed. By adding a little water, the lime would sizzle, generate heat, and turn hard as a rock.

As simple as this sounds, it is true alchemy. By scorching a rock you can re-create it anywhere and in virtually any shape imaginable. Call it rock on demand.

This initial discovery would take mankind down a dirt road of ramshackle shelters to a paved street lined with the biggest, strongest, and tallest structures.

Unfortunately, from its early beginnings, concrete had its downsides. Upon the discovery of lime, our ancestors invented a system to reproduce it, supposedly humanity's first industrial process. They extracted lime from quarries. They erected limekilns. Their axes chopped down large swaths of forests. They burned the wood in the ovens to heat the lime. But limestone consists primarily of calcium carbonate. This material is often the result of the skeletal remains of marine organisms, such as coral and shelled organisms. When it is heated, it releases carbon dioxide into the atmosphere. Burning wood also releases carbon dioxide.

The environmental toll continues today. Bulldozers carve away hillsides. Cement-factory chimneys pump out carbon dioxide, wafting into the atmosphere. Pipes extract tons of water to create concrete, the world's tenth most water-intensive industrial user.

The Greek myth of Prometheus is often used to highlight the double-edged sword of technology, and it could not be more relevant for concrete. Prometheus stole fire from the gods and gifted it to humans to fuel their progress. As punishment for his crime, the gods chained him to a rock where an eagle ate his liver, every day, anew. To counteract the blessing of fire, the gods created Pandora and gave her a mysterious box. She took off the lid of her box, unleashing evils and disease to infect the earth. It could be said that the lightning strike that gifted us concrete, the building block of modernity, also opened a Pandora's box of its own.

Initially, concrete played sidekick to traditional building materials like stone. The Egyptians built the Great Pyramid of Giza of stone blocks. They used lime only as cement, or "almost concrete," to create the joints in between the stone. It wasn't until the Romans that lime was used as an ingredient for concrete and became a building material in its own right. In fact, this material enabled the Roman Empire's lasting legacy. The Romans built coliseums and viaducts that lasted for centuries. The Romans coined the word "concretus," meaning "grown together," and gave us modern concrete.

The first Roman concrete was found near the slopes of Mount Vesuvius, just west of Naples, in the Italian town called Pozzuoli. Its name is derived from the Latin for "mineral springs," for the town's sulfureous pools with supposedly healing benefits. While Roman slaves dug deep pits to quarry limestone, they also dug sand to mix with lime and make concrete. But it is likely the Romans first mistook the local reddish-colored earth near the volcano for sand. They

added it to lime and found that the resulting concrete was exceptionally strong.

In reality, the reddish earth was volcanic ash. Roman engineers quickly realized that the chemical process between volcanic
ash and lime created a super mortar. Thanks to the alumina and
silica in the volcanic dust, the concrete was stronger, harder, and
cured faster than anything else they had seen. It could even harden
underwater. The Romans, knowing that the Pozzuoli's sand was
special, gave it a name: *pulvis puteolanus*, or "dust of Puteoli," now
known as pozzolana.

Seneca, the Roman Stoic philosopher, later said the "dust at
Puteoli becomes stone if it touches water." Vitruvius, the Roman
engineer and architect, wrote about the powder's "astonishing
results.... This substance, when mixed with lime and rubble, not
only lends strength to buildings of other kinds, but even when piers
of it are constructed in the sea, they set hard under water."[2] Pozzolana quickly became a staple in concrete's mix. Vitruvius's recipe:
one part lime to three parts pozzolana for cement used in buildings,
or one part lime to two parts pozzolana for underwater work.

Unlike modern concrete, which crumbles as it ages, Roman concrete became stronger over time. We now know this about Roman
underwater structures, such as breakwaters, which are stronger
than ever today, despite the harsh saline environments. In contrast,
modern concrete degrades within decades under the pressure of the
pounding salty waves. How is this possible?

Tobermorite. This rare crystal with a Superman-like name
and characteristics exists in Roman concrete as a result of a reaction between the volcanic ash and the seawater. The crystal makes
the concrete flex, rather than shatter under pressure, rendering
it extremely durable. With each new tiny crack as a result of the

waves, seawater seeps in and creates even more tobermorite. For this reason, one researcher named Roman concrete "the most durable building material in human history, and I say that as an engineer not prone to hyperbole."[3] Today, to find solutions to our deteriorating modern concrete infrastructure, geologists are X-raying two-thousand-year-old samples from Roman monuments.

The Romans not only engineered stronger and more durable concrete, they also used it to build entirely new, previously unimaginable structures. The Pantheon in Rome was arguably their most impressive architectural achievement, thanks to its 142-foot-wide lifted concrete dome, the largest of the ancient world, bigger than anything made from stone. Two thousand years and several barbarian invasions later, it continues to be the world's largest unreinforced dome.

We think of concrete as rock. But the Romans realized it is actually more like plastic. Unlike rock, where sculptors must chip and carve away piece by piece, it takes little effort to create almost any shape in concrete. In its liquid state, concrete can be poured into a mold, just like plastic. After a few days, the concrete hardens into a single solid. This is what the Romans found as they poured the Pantheon's cupola. They first constructed a wooden mold in the shape of a half sphere. They then poured the concrete mix over the mold, which had coffered-shaped recesses to create a lighter and vaulted ceiling. They left the peak of the sphere open to create an oculus, a circular shape to let in light. Armed with concrete, Romans created a vaulted dome so strong and so beautiful that it led to the building's name, "pantheon," "temple of all the gods" or heaven on earth.

Rock is a natural material. Concrete is artificial, a product of human imagination, just like plastic. The Pantheon's artificial stone dome symbolized a new level of humanity's manipulation of the natural world. As slaves quarried for rock to create concrete, humanity chipped away the natural earth to create artificial all-encompassing environments. The quest for this ultimate human-created environment, one fashioned entirely after our whims, would culminate in supertall skyscrapers. A stack of hundreds of entirely manufactured environments, each floor on top of another, each supplied with its own artificial light and conditioned air.

As the western part of the Roman Empire fell in the fifth century, so did our use of concrete. During the Dark Ages, for 1,300 years, concrete was relegated to being a mere infill material for walls or foundations made of natural stone. It wasn't until the eighteenth century when concrete's true potential was rediscovered by John Smeaton, an English civil engineer. He needed a stronger mortar for a new lighthouse on the treacherous Eddystone Rocks, located fourteen miles out at sea in the English Channel, which had burned down. Smeaton wanted a mortar able to set underwater. Learning from the Romans, he brought back Italian volcanic ash. Smeaton published his work, and concrete came back into fashion.

Builders, engineers, and inventors soon tinkered with new concrete-mix recipes. One such tinkerer was Joseph Aspdin, the eldest son of a bricklayer in Yorkshire. His 1824 patent, an "Improvement in the Mode of Producing an Artificial Stone," was of questionable originality. He named it "Portland cement," after the prestigious natural stone that was quarried on the Isle of Portland, and the name stuck. Today, Portland cement is the most common type of cement with an industry the size of hundreds of billions of dollars.

The key characteristic of Portland cement actually refers to a later invention by Aspdin's son, William. William found that when cement's raw materials are accidentally overheated, it creates hard balls the size of marbles, called "clinkers." Initially, people believed these clinkers were useless. But when ground into powder with millstones, these clinkers produce an extremely fine cement superior to anything else. Overheating cement's ingredients vitrified them into a single glassy substance in which the various molecule particles fused together into stronger aggregates, silica molecule structures. This process is called clinkering.

Unfortunately, clinkering also contributes to most of the carbon emissions of Portland cement production. Romans baked limestone at 1,652 degrees Fahrenheit (900°C). In contrast, Portland cement requires heating up to 2,642 degrees Fahrenheit (1,450°C), almost a thousand degrees higher, leading to tons of fossil fuel wasted. Then there is the problem of grinding the hard clinkers, which quickly wears the millstones down. In addition, Portland cement uses much more lime per unit of concrete, making it even more carbon-intensive.

A fundamental weakness remained. Concrete is very good at absorbing compressive forces, such as the force of gravity pushing down on a material, compressing it. However, it is bad at resisting the tensile forces that pull materials apart, such as from the shaking of a blowing wind or an earthquake. The Romans had tried to overcome this limitation by affixing bronze strips to the concrete, supplementing the compressive strength of concrete with the tensile strength of metals. But they ran into a problem. Bronze expands more than concrete as it warms. When the temperature changed, the strips of bronze pulled the slower-expanding concrete apart.

The idea was a good one. Reinforcing concrete with metal, a property good at withstanding tensile forces, would create a composite material, a combination of two different materials. The two materials together give the composite unique properties. For instance, combining mud and straw in mud structures creates a composite stronger than mud or straw by itself. Dried mud is good at resisting squeezing. Straw is good at preventing tearing. Together, the composite material makes for a stronger material. But by itself, straw crumbles when you squeeze it, and dried mud pulverizes when you tear it apart.

Given the Roman failure with bronze, the idea of reinforcing concrete with metal was abandoned until French gardener Joseph Monier in 1849 wanted to create a better flowerpot. He had grown dissatisfied with existing materials. Clay pots cracked easily. Wood pots rotted and were easy targets for penetrating plant roots. Concrete pots could work, as long as they could absorb tension. Monier decided to embed an iron mesh inside the concrete, and reinforced concrete was born. The love affair between concrete and steel had begun.

It was the best of both worlds. Concrete is strong in compression, but weak in tension. Steel bars are strong in tension, but weak in compression. In addition to their physical properties, the material combination also made practical sense. Concrete is relatively cheap, and is easily poured into almost any shape, so it is a good material to add bulk. Steel is more expensive, so best used sparingly, and is easily found in linear shapes, such as bars. Unlike bronze, steel has a similar coefficient of thermal expansion as concrete, so there would be no risk of cracks. Monier displayed his composite material at the International Exposition of 1867 in Paris, and the idea quickly spread. But a problem remained: how to better affix the steel bars to the concrete so that the two materials would act together.

A solution came over a decade later across the world. In 1884, the English-born Ernest Ransome was residing in California. Years earlier, his father, Frederick, had invented the rotary kiln, a 26-foot (8-meter) long revolving cylinder with a flame to make concrete production cheaper and better than in vertical-shaft kilns, thus ensuring concrete's commercial success. His son Ernest was an inventor in his own right. After performing experiments for reinforced concrete sidewalks, he patented twisted square "rebar" (short for "reinforcing bar"), instead of keeping the iron piece smooth. Initially, engineers ridiculed his idea, thinking the twisting would weaken the iron. But in reality the structures were stronger. The twisted ribs bound better to the concrete than the smooth bars did.

Today, the principle of increasing friction between the bar and the concrete remains the same—though we don't use twisted rebar, but ribbed rebar. Workers first place elaborate webs of thin, ribbed metal bars. Then, concrete is poured over the webs until they are entirely covered. A few days later, the concrete is fully cured and creates a new, composite material—reinforced concrete.

Ransome fought skepticism, and reinforced concrete had a long way to go before being accepted as the staple of construction. It took a tragedy to really change the perception of the material, and to vindicate Ransome's belief in its widespread use. A fire raged in the Pacific Coast Borax Refinery in 1902, a concrete-framed structure built by Ransome. The blaze was so hot it melted metal, but the concrete frame suffered only minor damage. Ransome demonstrated that concrete's resistance to fire was far superior to the dominant steel-framed buildings. Fire reduces metal's strength quickly, but concrete covers and protects the metal. Paradoxically, this fire was the triumph of Ransome's work,[4] propelling his career and making reinforced concrete seem a miraculous material.

The following year the world's first reinforced concrete skyscraper arrived. The Ingalls Building, completed in Cincinnati in 1903, gained its strength from Ransome's system, with twisted steel bars cast inside the concrete. The builders took two years to convince city officials that the 16-story tower would hold up. Skeptics in the building industry feared that a concrete tower so tall would collapse by the wind or under its own weight. Allegedly, a reporter stayed awake the entire night after the building's opening to get the scoop on the structure's downfall.

The reporter waited in vain. Instead, the building's architect, A. O. Elzner, published the findings, praising concrete's benefits over steel, not least of all for being "considerably cheaper." He noted, "Steel requires a great amount of capital and equipment and money to operate a steel plant. Long hauls and heavy freight bills are also involved."[5] Besides cost savings and better fire resistance than steel, solid and heavy concrete slabs transmit little noise, making them especially suitable for residential construction.

Still, it would take until the end of the century before reinforced concrete became the most common structural material for the world's tallest towers. Initially, it was used to strengthen the expressive shapes of several iconic structures. In the early part of the century, the Catalan architect Antoni Gaudí proposed using concrete to build the vaults of the nave of the Sagrada Familia, Barcelona's landmark cathedral, understanding that the lateral loads would be absorbed by the reinforcements in the concrete. In 1935, Frank Lloyd Wright used reinforced concrete to create Fallingwater in Pennsylvania, a villa that cantilevers far out over a waterfall, an impossible feat without rebar. The Italian engineer Pier Luigi Nervi built stadiums the shape of parabolas, such as his stadium for Rome in 1957, which has a reinforced concrete thin-shell dome spanning 194 feet in diameter. In 1964, Nervi also showed that reinforced concrete could reach new heights for towers, like the 620-foot (190-meter) Tour de La Bourse in Montreal.

Reinforced concrete and clinkering represented the most significant improvements to concrete in the nineteenth century. Meanwhile, significant innovations improved reinforced concrete in the twentieth century. Air entrainment, the addition of billions of microscopic air bubbles into the concrete mix, helped reduce concrete's weight by 3 to 9 percent. This process is especially useful in cold climates, since the air bubbles trapped in the concrete act as chambers for freezing water to expand into, enhancing concrete's resistance to freezing. At the same time, it improves the workability of the mix, allowing for fewer parts of water, which reduces concrete's strength.

In addition to air, fly ash made it into the concrete mix as well. This pulverized fuel ash, a by-product of burning coal, is so featherlight that it flies up the power plant's smokestacks and contributes to smog. The 1970s environmental laws forced coal-fired power stations to capture this ash. The industry found the perfect use in concrete. Not only can this siliceous material replace up to 30 percent of Portland cement, limiting the use of the most carbon-intensive material of concrete, it also increases concrete's strength and durability.

Finally, two new materials, superplasticizer and silica fume, helped create a "high-performance" concrete with superior durability and strength. A superplasticizer is a synthetic polymer that helps slow down the clustering by cement particles, keeping it workable longer. It also allows the use of up to 30 percent less water, making concrete stronger. And since the cement particles float around more before they bond, they can coat more of the aggregate of sand and stone, making concrete even stronger.

Silica fume, an ultrafine powder, helps concrete stay liquid longer as well. In some ways, this modern powder is a throwback to ancient Roman concrete. The silica reacts with the water and cement to create a pozzolanic reaction. When concrete cures, the same hydration reaction that creates calcium silica hydrate, the glue binding everything together, also creates calcium hydroxide, a junk substance that actually works against concrete's strength. A pozzolanic reaction converts some of this troublesome calcium hydroxide into calcium silica hydrate. Two thousand years after the Romans first discovered it, the strength of pozzolana persists.

Where Roman concrete had a compressive strength of 20 MPa,[6] high-performance-strength concrete can get up to 200 MPa. Despite this advancement in terms of strength, modern concrete was a step back in durability. The Roman viaducts still stand today in good condition. In contrast, the infrastructure built in the 1950s and 1960s has an average life span of fifty years, and is now up for costly repairs.

Sadly, the marriage between concrete and metal would be subject to slow and inner deterioration, due to iron's tendency to rust. Though concrete covers the iron and protects it from weathering, moisture still enters through tiny cracks in the concrete, triggering an electrochemical reaction between the oxygen, moisture, and metal on an atomic level. This creates iron oxide, or rust.

Rust not only weakens the bonds of the metal, it can also damage the concrete. Since iron oxide is a larger molecule than iron, the iron rod's oxidized diameter can expand up to four times its size, enlarging cracks and fracturing concrete. This process is called "spalling," otherwise known as "concrete cancer." Spalling is the cause behind the mass deterioration of today's bridges, especially those built in the mid-twentieth century, before more measures were taken to prevent rust. Without repair, they are bound for collapse.

Remedies exist, but they are costly. High-alloy steel, or stainless steel, contains a material that reacts with oxygen before steel does, such as chromium. This then forms a thin and stable film as a barrier against oxygen and water reaching the underlying metal. But it's up to ten times as expensive as normal steel. Zinc can also be used, which corrodes before the steel does. After the zinc is "sacrificed," the steel is next, leading to reinforced concrete's fate of eventual decay.

In 1972, the company 3M introduced an epoxy coating that made rebar more corrosion-resistant; still, unlike Roman concrete, modern concrete can last only about one hundred years without repairs.

Nevertheless, with modern concrete's superior strength, mankind's tallest structures reached new heights. In the early twentieth century, the two tallest towers, the Empire State Building and the Chrysler Building, were steel-framed buildings. In 1998 they were surpassed by the Petronas Towers in Kuala Lumpur, which were built mostly of reinforced concrete. The Taipei 101, also built predominantly of reinforced concrete, exceeded this record in 2004 with another few hundred feet.

Then the Burj Khalifa arrived.

Petronas Towers, Kuala Lumpur, César Pelli, 1999

I VISITED THE BURJ KHALIFA in 2010, the year it opened. It was my first time in Dubai and the city reminded me of Las Vegas, that other desert city that attracts millions of visitors despite its arid location. These cities, sisters in the sand, overcame their adverse geography by recreating fantasies in wastelands.

Dubai's efforts paid off. The city's extravagant building program generated its top three tourist attractions. Next door to the Burj stands the Dubai Fountain, featuring the world's largest choreographed fountain system inside a 30-acre artificial lake. WET design, the company that designed the Bellagio fountains in Las Vegas, engineered a system able to spray 22,000 gallons of water up into the air, the equivalent of what five hundred bathtubs can hold. These two attractions are magnets for the Dubai Mall, the economic engine of it all and itself the third most popular attraction, housing an Olympic-sized ice rink. Of course, ranking number one is the towering marvel, the Burj Khalifa.

None of these—the world's tallest tower, fountains, and an ice rink—are things you'd ordinarily expect to see in the desert. But now that they are there, they seem to make for a fitting achievement for a city built on superlatives: the biggest with the tallest and the mostest.

On the 143rd floor of the Burj Khalifa: the world's highest nightclub. On floor 148: the world's highest observation deck. When I reached the tower, the structure loomed so tall I could not fit it in a single photo— the world's only building that defies your camera lens, I thought.

This extravagance may seem foreign and out of scale, but that is precisely why it draws people and money to the desert. However, this game of tourism and real estate comes with seemingly glaring geographic contradictions. While hillsides are being erased to carve out limestone, the world's heaviest concrete building emerges on weak bedrock and silky sand. Ignoring water scarcity in the desert region, the tower consumes 250,000 gallons of water every single day.

The city of Dubai is an environmental paradox of its own. The Burj converted sky into stacks of floor space. The World, an archipelago of artificial islands in the shape of the world map, buried coral reefs in 450 million tons of sand for precious waterfront real estate. While water is poured on the desert to create artificial lakes, marine sand is dredged to turn water into land. These twin materials, dredged sand and concrete, underpin Dubai's extreme colonization of the earth and the sky. Both are slushy and viscous, and each alone is so remarkably behaviorally complex that they occupy the entire professional lives of hordes of civil engineers.

Ironically, the builders of the Burj needed to import sand from Australia, despite the region's abundance of sand. Wind bounces desert sand along the surface, making the grains too small and too round to be useful for building. Marine sand, in contrast, is larger and more angular.

Upon completion, the Burj Khalifa reveals few traces of its half-mile-long concrete structure, the skeleton underneath the skin of shiny glass and metal mullions. There is no sign of another skeleton in the closet as well—the sweat and blood of the underpaid Indian migrant labor that built it. Few critics pointed to the construction workers' low wages, crowded and unsanitary dormitories, and confiscated passports until their completion of duties. Draconian sponsorship laws push many migrants into despair, with hundreds of suicides a year among Indian migrants alone across the emirates, and suicide rates of expatriates that are seven times higher than the national population.[7]

An abundance of money, of course, also made it possible. When the Dubai developer ran out, its oil-rich neighbor Abu Dhabi chipped in. It led to an unexpected last-minute name change. What was initially called Burj Dubai became Burj Khalifa, after the ruler of Abu Dhabi, Khalifa bin Zayed Al Nahyan.

The design talent was sourced from Chicago, where the architecture firm of Skidmore, Owings & Merrill (SOM) had built a reputation for the design of very tall buildings. Chief architect Adrian Smith paid tribute to the Middle Eastern architectural context, with gradual setbacks reminiscent of the spiral minaret of the Great Mosque of Samarra. Yet, even with all the design, money, and political will afforded a wealthy country, the Burj Khalifa would not have been possible without concrete.

Besides concrete's strength, it offered massive logistical benefits for a skyscraper the size of the Burj Khalifa. Hauling stones skyward, one by one, to build such a skyscraper would have been nearly impossible. The genius of concrete is that, despite its ultimate strength, in its liquid state it can be pumped through a pipe. Burj's success was entirely dependent on its ability to pump. Placing concrete by crane by lifting cement buckets would not only have been prohibitively expensive, but it would also have bogged down the whole process.

Never before had concrete been pumped above 1,600 feet. Many things could have gone wrong, and the heat of the desert location added to the challenge. As the concrete gushes up the pipe, heat can harden it before it reaches its destination. A blocked pipe would mean the failure of the entire project.

To solve this issue, the German chemical producer BASF created an admixture called Glenium Sky 504, a superplasticizer that keeps the mix soft for a full three hours upon arrival. The concrete would have to be poured at night. Daytime construction in the summer months, with temperatures reaching up to 122 degrees Fahrenheit (50°C), is too hot. Flake ice would keep the mixture cool, at a maximum of 90 degrees Fahrenheit. Yes, this ice-chilled concrete was a sophisticated blend.

However, even with this delicate chemical concoction, the concrete mix may wear the pipe down. Concrete consists of aggregate, not just fine cement, and its coarse little lumps scrape the inner pipe. A pipe can pump only about 10,000 cubic meters of high-abrasive aggregate,[8] about as much as four Olympic pools, before it breaks down. This was a problem in the case of the Burj when there was an equivalent of 130 Olympic pools that had to be pumped. The higher the concrete mix needs to go, the higher the pressure, which also means more damage to the pipe. Plus, the coarser the aggregate, the faster the wear. A large aggregate size helps reduce the necessary water, with a lower surface area to coat as compared to small coarse aggregate, enhancing the strength of concrete. It seemed like a catch-22.

To find the ideal diameter of the pipe, the size of the aggregate, and the pressure needed to pump the concrete, the team could not rely on precedents. They needed to understand the mix's friction and flow. A pumping trial began. Engineers laid out 600 meters of high-pressure delivery pipe in bright sunlight and added sensors to measure the concrete's pressure. They eventually decided on a 150 mm diameter pipe, allowing for a maximum 20 mm aggregate, which gave each pipe a life span of 40,000 cubic meters. To minimize the abrasion damage when pumping to the top, where pressure needs to be higher, they reduced the aggregate above 1,135 feet from 20 mm to 14 mm.

With these pipe and concrete-mix calibrations, the engineers also needed a pump with a high enough pressure. This is where the Putzmeister came in. Putzmeister (aptly named "Plaster master" in

German) is a German manufacturer of concrete pumps and a world record holder for volume of concrete pumped. During the Chernobyl nuclear accident in 1986, their pumps poured 400,000 cubic yards of concrete to entomb reactor four. With the nuclear accidents at the Fukushima plant in 2011, once again Putzmeister concrete pumps were flown in.

For the Burj Khalifa, the company created the Putzmeister BSA 14000 SHP-D, a new high-pressure trailer concrete pump. Its hydraulic pressure tops out at 31 MPa, even more pressure than a high-power bullet striking a bulletproof object. The company brought over a total of three pumps, each connected to their own high-pressure pipes, so that concrete could be placed at three locations simultaneously. Putzmeister managed to pump the concrete up to 1,972 feet, shattering the record of the Taipei 101 tower's 1,542 feet.

As the concrete reached the top, it still needed to take shape and cure. Tons of metal and plastic formwork were shipped from companies around the world.[9] The next challenge was to reuse the formwork, and to minimize the "floor cycle," the number of days it takes to place one floor. To place the core walls in the shortest possible time, for instance, required thoughtful construction coordination and "self-climbing" formwork weighing about 1,000 tons. On day one, a preassembled cage is installed, which is inspected the next day. On day three, the concrete is cast. Ten hours later, the form is hydraulically jacked up to the next floor.[10] After trial and error, this process enabled a three-day-per-slab cycle. While not as fast as New York's speedy two-day cycle for a typical high-rise building, for a tower its size it represented a record, breaking the 4.3-day cycle of the Petronas Towers built in Kuala Lumpur.

This is how concrete made its way to the top. It was the most amazing part of the building, but was outshined by the flash and glitz. Yet this global and historic effort required an accumulation of inventions from all over the world—Roman engineering, American rebar, and a German pump—all in the Arabian Desert.

HUMANS ARE NO strangers to shortsightedness in building. Easter Island was once a thriving civilization, until the natives wasted natural resources on creating enormous statues. Several anthropologists believe the Easter Islanders cleared trees to make logs to roll the statues. One tree at a time, they eventually wiped the island clean. The deforestation caused erosion, which led to food shortages, creating conflict, internal violence, and eventually, societal collapse.

"Many of the great ruins that grace the deserts and jungles of the earth are monuments to progress traps, the headstones of civilizations which fell victim to their own success," writes Ronald Wright in *A Short History of Progress*. Natural depletion also explains the fallen civilization of Sumer, in Mesopotamia, considered the cradle of civilization, as well as the fatal undermining of the Mayans and the Khmer. Of the monuments in our own time, the Burj Khalifa is the tallest of all. Although a sign of progress, it may equally be a trap, a concrete monolith contributing to environmental decay.

We are building these giant buildings out of concrete. They soar beautifully into the sky, their graceful figures piercing the clouds. But back on earth, they are leaving an immense environmental footprint. They are not built to last. Why are we building the world's tallest buildings and vital infrastructure with short-term thinking? If the Pantheon would be built today with our concrete instead of the Romans', it would have to be rebuilt dozens of times.

Even the Burj Khalifa was designed to only a 100-year life service. The highly corrosive groundwater in the desert will gradually eat away at the reinforced concrete. With stainless-steel rebar, reinforced concrete's life span may be doubled, but eventually it is still bound to lose its structural integrity. That's a waste not only of concrete but also of an enormous amount of rebar. The 31,400 metric tons of rebar used in the Burj would cover a quarter of the world's circumference, if laid around the planet. Even with rebar's average recycling rate of 71 percent, as estimated by the Steel Recycling Institute, the amount of rebar destined to end up as landfill would span from New York to Texas.

There is another problem with concrete. It creeps. Creep is usually not a problem since it tends to be as small as one-eighth of an inch per floor over fifty years. But when a building is 160 floors, all those tiny bits add up.[11] Fifty years from now, it is likely that the Burj will be two feet shorter. This could wreak havoc for all the cables, pipelines, and ducts that could be subject to breaking.

The Burj Khalifa may be an easy target for environmentalists, but the tower is emblematic of a larger, systemic problem. Corrosion is eating away our buildings from the inside, responsible for most damage to reinforced concrete structures. When essential infrastructure such as bridges consists of reinforced concrete, they require regular testing to detect corrosion and prevent collapse.

Governments spend billions of dollars every year in concrete testing and repair costs. Often, these costs are not considered when concrete is specified. In the United States, it was calculated that the cost of corrosion could add up to 5 percent of the nation's GDP.[12] The true cost of reinforced concrete is much higher than most realize.

Rapidly modernizing countries will be up for a retroactive bill as well. In 1980, the year after China opened up its economy, it produced less than 80 megatons of cement. By 2010 this had increased by nearly 24 times to 1.9 gigatons, which represented 56 percent of the world cement production.[13] Then, in the next three years, it used a total of 6.6 gigatons of cement, more than what the United States used in the entire twentieth century—the century of the Hoover Dam and the Interstate Highway System.

Needless to say, all this concrete has contributed to people's quality of life. Dirt roads were paved with concrete so people could travel more safely and efficiently. Mud floors were replaced with concrete to improve sanitation. Concrete dams were built to protect from floods and to generate hydroelectric energy. But today, China has 23,841 large dams across the country, and many need serious repair.

Cement production is one of the most carbon-polluting industries. How can we reduce this? We can either build less, or build wiser. Since we're unlikely to be willing to build less, then we should rethink concrete.

Thanks to environmental regulations, concrete recycling is becoming more common. Crushing machines are used increasingly at demolition sites, where metal "jaws" shatter the concrete like a nutcracker, and magnets remove rebar. The aggregate then finds itself in new uses, such as in a sub-base gravel at the bottom of a road.

Some companies are looking for a better way to manufacture cement. The company Heliogen has a cement plant with a tower surrounded by hundreds of mirrors out in the Mojave Desert. Each of the mirrors points at a steam turbine inside the tower, intensely focusing the sun beams, fueling the cement factory with solar power instead of gas or coal. As promising as it may be, the factory is contingent on areas of ample sunshine.

Could we do without concrete altogether? If we were to use

steel, the total carbon use would be even worse, since steel has a much higher carbon footprint. We need a new concrete that can last and does not damage the environment.

Already, a quiet war on concrete is being fought. AshCrete is a concrete alternative that replaces up to 97 percent of traditional cement with fly ash, a recycled material made of the ash waste from coal-fired power stations. Unfortunately, it relies on the burning of coal and it's more susceptible to cracking when it freezes and thaws.

Timbercrete uses sawdust from sawmills and mixes it with concrete, making it twice as light as concrete and a better insulator. It can be molded into shapes such as blocks and bricks, but its compressive strength is nowhere near enough for tall skyscrapers.

Some propose materials to be grown, such as a brick made of mushrooms, and entirely without cement. By leaving mycelium, a part of the fungus, and crop waste such as corn husk in a mold to grow for five days, a biodegradable brick emerges with an otherworldly aesthetic. Unfortunately, this brick is strong enough for structures of only a few stories.

Scientists at the University of Colorado, Boulder, are using cyanobacteria, tiny organisms known as blue-green algae, to grow bricks. The microorganisms consume carbon dioxide to transform gelatin and sand into a bacteria-filled brick that includes calcium carbonate, an ingredient in limestone.

But the living brick is weaker than conventional concrete. Meanwhile, concrete may become home to other living organisms as well, such as bacteria or fungi that can automatically repair cracks, creating a material called "self-healing" concrete.[14] Concrete can also be made more fracture-resistant, or even bendable, through nanoengineering. Ductile concrete, so-called Engineered Cementitious Composite (ESS), is inspired by nacre, the iridescent coating of abalone shells. A unique combination of ductility and strength protects the abalone from sea otters that toil to break the brittle shell by pounding it on rocks.[15] Bendable concrete includes internal fibers about 10 millimeters long and 40 microns in diameter, roughly half the thickness of a human hair. These thin polystyrene particles bridge micro cracks into the concrete, allowing the material to absorb energy and increase its tolerance to damage, which is useful in high seismic areas. Unfortunately, it has an initial cost up to four times higher than traditional concrete.

These technological frontiers, ranging from nanotechnology to synthetic biology, could suggest exciting new ways to avoid the problems with modern concrete. But it's still hard to top Portland cement, which has been in use for two centuries, despite its high environmental costs. The construction industry tends to lag behind other industries in terms of innovation. Restrictive building codes, conventional building practices, and general risk aversion lead to slow adoption of new technology.

But there are also ways to replace concrete with another building block. The most promising substitute for concrete in buildings is a throwback to our most ancient construction material: mass timber. Cross-laminated timber (CLT), is an engineered wood panel consisting of glued layers of lumber board, each oriented perpendicular to the layer above. The resulting product has an alternating grain, making it stronger than conventional lumber, which is weak against the load perpendicular to the grain. Not only does it require less energy to produce than concrete, it also sequesters carbon. When wood is taken out of the forest, trees planted to replace the harvested ones can sequester new carbon. Wood is a carbon sink, not a carbon source.

A recent study shows that, on average, a high-rise construction of mass timber emits a quarter fewer carbon emissions than concrete, and this may be a conservative estimate.[16] It is also a biodegradable material. Nevertheless, because cross-laminated timber requires adhesives, it can still end up as waste.

As long as the forests are sustainably managed, mass timber could be a building block for lower-carbon structures, and one with multiple benefits. Timber can be easily cut in several shapes and sizes using computer-directed cutting machines. In contrast, concrete must be cast in a mold, and is more time-intensive. Since it's so light, timber is easier to move to a site and operate. However, it requires standardized parts optimized for manufacturing and assembly, much like interlocking prefabricated elements, such as LEGO. Or think about IKEA. Instead of assembling a dresser, you get a flatpack skyscraper. This method may also allow the structure to be disassembled more easily at the end of the building's life.

Surprising to many, mass timber performs well in a fire, unlike small-sized lumber, because the char forms a protective barrier that prevents the fire from reaching the interior. Recently, code provisions have been approved for the 2021 International Building Code to allow mass timber for buildings up to 18 stories.

Several countries already have "plyscrapers" this size, with larger ones to come. In 2017, a 170-foot-tall student residence was completed at the University of British Columbia in Vancouver. It was created about four months faster than typical buildings of similar dimensions, thanks to the prefabricated elements and the light weight. In 2019, Norway overtook the record for the tallest mass timber building with a 280-foot-tall building in the town of Brumund-dal, which includes a hotel, restaurants, offices, and apartments.

The success of these structures may increase interest in taller timber towers. This will require more strength. One possible solution is combining mass timber with reinforced concrete, creating a new concrete-wood hybrid system. The architectural firm SOM teamed up with Oregon State University to test a hybrid floor system—a 9-inch-thick CLT panel topped with a little over 2 inches of concrete. The floor was able to support 82,000 pounds, eight times more than required. While this hybrid system can be designed for 30-story buildings, there is still a long way to go before supertall status.[17]

Mjøstårnet, Brumunddal,
Voll Arkitekter, 2019

Concrete is both a blessing and a curse. The mysterious gift of liquid rock helped us build society's largest and most marvelous structures. But it also contributed to climate change. Can we reduce our addiction to concrete? If we want to keep building, we need new recipes, new technologies, and new alternatives that improve on concrete. Our future may depend on it.

TWO

The Fight Against Sway: Wind

Even before the Burj Khalifa topped out in Dubai and claimed the title as the world's tallest building, on the other side of the Arabian Desert, the House of Saud responded. In 2008, they announced a mile-high tower (1.6 km), twice the size of the Burj, to be called Mile High Tower. Half a century after Frank Lloyd Wright's proposal for a mile-high skyscraper, would his vision finally become a reality? But a structure that size would draw a lot of wind, especially at higher altitudes where wind blows faster. Even if the building could withstand these forces, its upper floors would sway. People might get too queasy to enjoy their mile-high view over the Red Sea when they unexpectedly see their martini stirred and shaken.

The Saudi skyscraper plan was soon downsized to a "mere" kilometer-tall building. This unfortunate imperial-to-metric translation was blamed on the unfavorable soil reports, which revealed the weak underlayers of limestone and sandstone, both porous rocks, unsuitable for a tower of such height. But since a "Kilometer High Tower" just didn't have the same ring to it, they renamed it the Jeddah Tower, after the city it was going to be located in, the

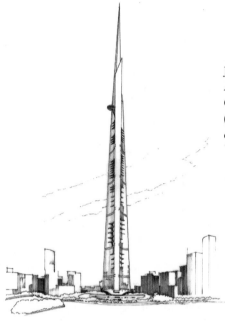

Jeddah Tower, Jeddah,
Adrian Smith + Gordon
Gill Architecture, 2009
(construction currently
on hold)

country's commercial center. At its new height, it would still be the
world's tallest structure, three times the size of the Eiffel Tower and
a good 500 feet taller than the Burj Khalifa.

The Jeddah Tower, on which construction has currently been
stalled, is intended to house a Four Seasons hotel, luxury condominiums, office space, and the world's highest observatory, with
views stretching 60 miles into the Arabian Desert. A thin steeple is
planned to form a sleek shape in the skyline. The only protrusion was
designed for floor 157, where a saucer-like sky terrace would pierce
out, the outdoor playground of a yet-to-be-priced but surely exor-
bitant luxury penthouse (it was initially meant as a helipad, until
helicopter pilots deemed it an unsuitable landing spot). There, glass
floors offer future visitors an abyss view of a whopping 2,000 feet.

As an added bonus, at this height, the ice in visitors' cocktails may not melt as quickly as down in the desert, with temperatures about 9 degrees Fahrenheit cooler than at the base.

But, first, the 1 km steeple would have to overcome substantial natural forces. Supertalls flirt more dangerously with Mother Nature than other buildings. Besides defying gravity, which pushes the building downward, a supertall also needs to overcome the force of the blowing wind, which pushes from the side. As a building doubles in height, its gravity load also doubles—but as the wind speed doubles, the wind pressure on the building *quadruples* (wind force increases with the square of the wind speed). On a typical day, wind can exert up to 17 pounds of force per square meter on a high-rise building—as heavy as a gust of falling bowling balls.[1] To make a bad situation worse, the shape of the steeple could cause the wind to accelerate into dangerous vortexes. Not only is nature a factor for the tower, the tower has a natural force of its own.

Strong materials alone, such as reinforced concrete, do not make a supertall. To fight the forces of nature, and especially the wind, a skyscraper's structure is vital. Even if a building supports itself through its aboveground structure, it still needs support from the ground, the substructure. Without a proper foundation, buildings this heavy would fall, sink, or lean—like the Leaning Tower of Pisa. Even with the structure able to keep the building standing, the building's top may sway in the wind, causing people to experience discomfort.

Now that we know how to create supremely strong materials, we still need to figure out how they come together in their best possible geometric shape. Mankind took thousands of years of experimentation to get to the Jeddah Tower's structure. Our ancestors stacked stones on stones, first creating simple walls, then Stonehenge-style portal frames. With sometimes deadly trial and error came arch shapes like the Roman bathhouses and the elaborate flying buttresses of Gothic cathedrals. But in the process, as our structures evolved, we were given vast interior spaces so beautiful, they seemed like heaven on earth.

Man's tangling with Mother Nature may get us in trouble. After all, her forces have been around for billions of years, and we've only just started pushing further into the sky.

THE ANATOMY OF A BUILDING is not unlike that of a human. The building's facade protects the interior from the weather, just like our skin is an organ of protection. Plumbing and ducts in buildings nourish occupants with water and air, the way our veins and arteries move vital resources around. The building's structure holds everything up, as the human skeleton does.

While the material of the structure is important, the structure's shape is equally essential. In addition to our bone tissue, the overall geometry of our skeleton keeps us standing. Our body stands up because our skeleton has a stable, two-legged base and a strong spine. Without our ribs, our lungs would collapse.

The human metaphor for buildings also rings true in another way. Buildings, like humans, progress as a result of Darwinian evolution of the fittest. Primates moved into treetops as their feet evolved into hands. So too did our buildings reach higher grounds as we invented new structures and evolved greater knowledge of building. Our Neolithic ancestors developed the first structural system with only three elements. Two stones stand upright as columns, holding a third stone on top, lying horizontally as a beam. The post and lintel was born. These portals could be combined into larger systems, such as Stonehenge, a ring of standing stones.

This most famous prehistoric monument is fourteen feet tall. Today, the buttressed core of the Jeddah Tower may bring us more than two hundred times higher, better able to fight gravity and the wind.

While basic, the portal geometry formed the root of architecture and represented a revolution for our way of living. In contrast to entirely solid structures, like some of the early Pyramids, the portal allowed for an opening into a space. Humans could use these cavities for habitation or for religious ceremonies. Instead of finding shelter in cavernous mountains, or digging in mounds, humans were now able to create strong and permanent protection anywhere—provided they had access to stone. Now humans could create their own caves.

The ancient Greeks developed the post and lintel system into an art form. Initially, they used wood to construct their temple frames. They carved their columns, logs of lumber, accentuating their vertical shapes with lines, while adding decorative bases and capitals. As they gradually replaced wood with the more permanent stone, the wooden details persisted. Expert carvers developed entire styles, such as the Corinthian, Doric, and Ionic capitals, the last the shape of spiral scrolls.. The lintel itself became a three-pieced element with an architrave, cornice, and frieze depicting mythical scenes. This system found its ultimate expression in the Parthenon, the temple complex on the Athenian Acropolis in Greece. Not only did the ancient Greek post and lintel represent the classic temple fronts; in the eighteenth century, it became a staple of neoclassical architecture and continues to be widely used today.

But post and lintel has its flaws. It allows for the broad protection of the elements, but only for a limited number of people. It can hold up only limited weight, constrained by the strength of the two individual columns. And it yields only limited space between the posts, constrained by the strength of the individual beam. For larger buildings, the system would lead to a sea of columns, which is not the most practical use of space.

It would take thousands of years for the next evolutionary step after the post and lintel.

Builders in Mesopotamia, a region between the rivers Tigris and Euphrates with a limited supply of stone and timber, used local clay to sunbake bricks. However, bricks, unlike stones, are too small to create a lintel opening. The Mesopotamians came up with a solution around 2,000 BC, changing architecture. They invented the arch.

The blocks in an arch, a semicircular structure of wedge shapes, support each other by their mutual pressure, distributing the pressure through the entire form of the arch. Therefore, they can withstand more weight than a single lintel stone. So-called fake arches existed earlier, such as corbel arches that spanned areas by overlapping stones, but they were not entirely self-supporting and so they were an evolutionary dead end. With sunbaked bricks in Mesopotamia, the true arch saw the light.

The limited strength of the sunbaked brick constrained the size of these arches, and hence the openings they could span. On top of this, some societies like the ancient Greek distrusted the arch. Since it always exerted a force outward, all stones needed to stay in check. Move out one of the stones, and the whole arch would collapse. It led to the old proverb: "The arch never sleeps."

The Romans perfected the arch, and as skilled as they were, they benefited from a few geographical advantages. History teaches us that evolution is predisposed to occur with the best local conditions. For instance, anthropologists explain the emergence of agriculture in the Middle East's Fertile Crescent as a result of the region's mild Mediterranean climate, abundance of crops, and high percentage of "selfers," plants that do not rely on other plants for reproduction. Likewise, the perfect arch sprouted in the Roman Empire because the Romans, unlike the Mesopotamians, had access to abundant stone, knowledge from Etruscan stone builders, and superior cement to bind the stones thanks to limestone and volcanic ash.

This allowed the Romans to create better arches and unprecedented structures. They typically built a wooden framework to temporarily support the stones as the arch was being formed. After they placed the center block, the keystone, they removed the frame. Once a line of arches was complete, this would become the foundation for another line of arches.

By stacking arch on arch, they created even bigger structures. This method allowed the spanning of ravines and valleys, with structures reaching up to 160 feet in height, such as the Pont du Gard, the aqueduct that still graces the Gardon River in southern France. Mankind, for the first time, was able to move water to places it couldn't before. It was a huge leap forward.

Romans could make even stronger and larger structures with concrete. Where the Pont du Gard was built out of soft limestone blocks, other aqueducts used brick as facing and then poured in concrete, such as at the Acueducto de los Milagros. The Roman Colosseum was built this way as well, the largest structure in Rome, at 159 feet, able to hold 50,000 people.

The Romans even elevated the arch to become a symbol of success. They built triumphal arches like the Arch of Titus, which features an elaborate relief inside the arch, showing the "Spoils of

Jerusalem." Arch shapes continued to be an inspiration for victorious occasions and assertions of state power throughout history,
from the nineteenth-century Arc de Triomphe in Paris to the 1965
Gateway Arch in St. Louis, Missouri.

The arch was just the beginning for Roman builders. They developed several derivatives of the arch, such as vaults. These became
structural innovations in their own right. The vault is essentially
an extruded arch, such as a tunnel vault, a continuous semicircular
arch spanning an entire space. Using concrete together with brick,
workers were able to build spaces spanning wider than before. This
led to basilicas and bathhouses, allowing Romans to hold large gath-
erings under a roof and to bathe in spectacular spaces.

The arch also spawned the dome, a hemisphere, which is
essentially an arch revolving around a central axis. Because of
its three-dimensional geometry, the dome allowed for even more
impressive architecture, like the Pantheon. Vaults and arches vary
in only two dimensions: height and depth. But the dome, unlike
the arch or vault, has a circular-shaped plan. It may also vary in
width. The dome offers spatial variations in all three dimensions
and presents a richer feast for the eyes.

After the Visigoths sacked Rome in the fifth century, it would
take centuries for further significant structural innovation. In
the twelfth century, the Catholic Church in France wanted larger
churches with larger window openings and larger spaces to house
more followers.

Abbot Suger was the first minister to Louis VI and Louis VII,
and he played a major role in the French monarchy. He felt that the
Church of Saint-Denis, where the monarchs were buried, was too
small to house the pilgrims. Instead of the heavy walls of Roman-
esque basilicas, he wanted slender columns with clerestory windows

to let in light. "The whole church would shine with the wonderful and uninterrupted light of most luminous windows, pervading the interior beauty," he wrote.[2]

In 1135, Suger summoned masons from several regions and started to rebuild the church. His enlargement of the choir, where he wanted vast windows containing unusually large stained glass, would reduce the structural stability of the walls. In addition, the taller he built the nave, the more lateral wind force the walls needed to support, and the thicker the exterior walls got. He needed something to absorb the horizontal thrust of the interior domes.

The flying buttress solved this problem. It was likely first applied by Suger's masons in Saint-Denis.[3] This outer buttress on the cathedral's exterior, consisting of an arch leading to a heavy vertical pier, transfers the side force from the inner nave into the ground. This redirects the outward push on the exterior walls in a more gradual, curved pathway to reach the foundation. Think of the building with buttresses as a person holding ski poles, and therefore more stable.

Not only does the flying buttress avoid very thick walls, it also helps create openings in the walls and bring in more light. "Once the new rear part is joined to the part in front, the church shines with its middle part brightened," wrote the abbot. "For bright is that which is brightly coupled with the bright, And bright is the noble edifice which is pervaded by the new light."[4]

Gothic architecture had arrived. Saint-Denis became the prototype for church design.

Fierce competition between dioceses led to ever larger structures. The bishop of Paris, Maurice de Sully, wanted to outdo Saint-Denis and its connections to the monarchy. He planned the largest cathedral of its kind, the Notre Dame, reaching 115 feet high and 427 feet long. With their forest of semi-arches, flying buttresses added to the landmark's aesthetic appeal. For a quintessential image of Gothic Paris, you only need to add a gargoyle spouting water out of its muzzle.

EVERYTHING CHANGED in the mid-nineteenth century with the mass production of steel. Already in the eighteenth century, cast-iron columns offered greater strength and a smaller circumference, reducing the mass of buildings and allowing larger spaces. When you mix iron with carbon to form steel, you get an even stronger and harder material. In addition, steam-powered cranes could more easily move steel beams around the construction site.

Since steel members are most easily produced in linear forms, structures initially returned to the simplicity of post and lintel frames. Although a steel frame may look similar to a stone post and lintel, it's entirely different structurally. Earlier post and lintel beams were each independent units because of their loose connection. Steel post and lintel frames, if they were properly bolted and welded together, became one unit structurally, with stresses distributed throughout. Another benefit was that steel frames could be erected more quickly than laying rows of bricks. Steel frames were significantly lighter than load-bearing masonry walls, allowing steel-frame portal structures to go even higher. For a good hundred years, steel portal structures became the preferred system for building skyscrapers.

Many define the first skyscrapers by their use of steel-frame construction. Chicago had the first with the completion of the ten-story Home Insurance Building in 1885. The steel used in the Home Insurance Building to support the building was only a third of the weight of a similarly sized building made of masonry.

Steel-frame buildings did not just lead to a different structure, they also changed the building's skin. The development of a load-bearing metal framework, independent of a facade, relieved the facade from carrying the weight of the building. With the facade liberated from carrying anything but its own weight, it could now be an entirely different material. Gone were the days of thick masonry walls. With their only objective to keep out the weather, entire walls of glass were even possible. This would lead to the standard sky-scraper, a steel frame with all-glass facades hanging from the slabs, the so-called curtain wall building.

When architects saw the potential of steel columns to transform buildings, the movement of modern architecture came to life. Swiss architect Le Corbusier began his Five Points of Architecture, the revered manifesto for modern architecture, with "Pilotis," French for "stilts." He claimed that the new grid of columns, instead of load-bearing walls, would form the basis of an entirely new aesthetic. Le Corbusier reveled in the opportunity to open up the ground floor and place the building on stilts.

In addition, instead of the small windows on typical masonry buildings, Le Corbusier wanted long horizontal bands of windows that would allow for expansive views. He applied his ideas initially on a small building scale, like Villa Savoye, a three-story building completed in 1931. Two decades later, he would codesign one of the first curtain wall skyscrapers: the 505-foot United Nations Secre-tariat Building in New York, an entirely glass-clad monolith.

It took a while for most skyscrapers to achieve this glassy look. Early curtain wall buildings encased their structural steel frame with concrete and terra-cotta, then filled in their facades mostly with brick, to help with fireproofing. In New York, this became the formula for ever taller buildings. The race for the world's tallest building had begun.

The Chrysler Building was the first to reach what we now consider supertall status, taller than 300 meters. In 1929, the builders were locked in a race with another skyscraper being constructed in New York, at 40 Wall Street. To win the battle, the builders had secretly constructed a spire pinnacle inside the building, so nobody would know the structure's ultimate height. As soon as 40 Wall Street had topped out, they lifted it up, "like a butterfly from its cocoon," recalled the architect.[5] The city and the builders behind 40 Wall Street were in shock.

The Chrysler Building's victory was short-lived, however. Within a year, it was overtaken by the Empire State Building, which was two hundred feet taller, reaching 1,250 feet.

Both the Chrysler Building and the Empire State Building are steel-frame skyscrapers. But using this structural system for a tall building came with a major downside. The steel columns and beams form a three-dimensional grid throughout the entire building. Since the column grids are closely spaced, there are virtually no column-free spaces on each floor. In addition, the towers used a riveted steel frame, which relies on intensive manual labor to rivet all the joints. With 391,881 rivets used in the Chrysler Building alone, it would be prohibitively expensive to build something like that today.

Back in Chicago, home of the first steel-frame skyscraper, a radical structural departure for tall buildings was brewing. The Bangladeshi-American engineer Fazlur Khan, engineer at the architectural firm SOM, invented a "tubular frame." Inspired by the structural principle behind bamboo, a hollow tube, from his hometown of Dhaka, he discarded the inner grid and returned the structure to the outside of the building. He decided that, with enough framing and with braces and trusses, the exterior could be the entire structure.

Khan wanted an exoskeleton, a building showing its structure. The term borrowed from the animal kingdom, referring to insects such as cockroaches, crustaceans such as lobsters, and shelled mollusks, like clams. All these species carry their skeleton on the exterior, in contrast to species such as humans, who are endoskeletons, carrying their skeleton on the inside. For these smaller species of animal, an external structure offers greater protection against predators, while helping support their body. For tall buildings, they protect the building from hostile wind forces, the dominant force on tall skyscrapers.

By putting the strongest part of the building on the outside, where the building stands at its widest, the structure has more stability. Imagine standing inside a shaky subway car. As you brace yourself for deceleration at the next stop, you may widen your stance to become more stable.

Khan first tried the tubular frame on the DeWitt Chestnut Apartments, a 42-story apartment building completed in Chicago in 1966. A grid-like facade of concrete columns and spandrels acts as a tubular structure absorbing all the horizontal forces. With the John Hancock Center, completed in 1969, Khan was able to expand his tubular frame to a whopping 100 stories, almost as tall as the Empire State Building. After an initial analysis, he realized that using a conventional structure for a building that tall would make it economically unfeasible. He had no choice but to innovate.

The exoskeleton structure had economic benefits. A typical steel-framed skyscraper required an internal cage of columns, which took up a lot of leasing space. The tubular frame moved the structure from the central area out to the perimeter, thus avoiding

John Hancock Center,
Chicago, SOM, 1969

a sea of columns throughout the office floor. Companies leasing the
space could now configure the floor any way they wanted.

To make the tower more stable, Khan added large "X"s, crossing
diagonals similar to a braced frame, which would make the building stiffer. Similar to the flying buttresses, these diagonal lines gave
the horizontal force of the wind an easier path down to the ground.

However, even within his own firm, doubts about the structural
feasibility of Khan's structural gymnastics persisted. To prove it
would hold up, Khan introduced computer models to calculate the
equations. Still, the firm decided to get outside experts, one of whom
suggested a new design by a different team. Khan threatened to quit.[6]

People also wondered whether the building would sway too
much at the top. Being inside a building this tall can be particularly

uncomfortable. Even with a strong structural shape, you could still find yourself swaying back and forth more than three feet on the top floors during a hurricane. A little movement is necessary, of course; in order to remain standing, buildings need to be both stable and flexible. While the wind should not push buildings too far to the side to the point of collapse, they do need to bend a little to absorb some of the wind energy in the structure. Especially in seismic zones, if the structure were entirely rigid, shaking might break the structure. Keep in mind there's a human limit to "good vibrations." While a little shake may not make a building topple, it may have some people at the top running for the bathroom sink.

On a fortuitous Sunday afternoon family outing, Khan and his family went to the Museum of Science and Industry in Chicago, where he saw a washing machine exhibit called "Tale of a Tub."[7] It featured a rotating platform meant to illustrate the vibrating motion of a Maytag washing machine. As he and his daughter stood on the rumbling tub, Khan had a breakthrough. He called the museum and modified the platform to stage a test. As volunteers stood and sat on the swaying platform, they annotated their perceptions. With this experiment, Khan demonstrated that the tower's upper occupants would stay comfortable. His design quickly got the green light.

Khan ultimately proved his skeptics wrong and left an enduring legacy on the Chicago skyline. With its dark angular structure dressed on the outside, the John Hancock tower gave a masculine appearance. The tower was praised worldwide. "Dark, strong, powerful, maybe even a little threatening—like a muscle-bound, Prohibition-era gangster clad in a tuxedo—the John Hancock Center says 'Chicago' as inimitably as the sunburst-like summit of the Chrysler Building evokes the jazzy theatricality of New York," wrote Blair Kamin, the architecture critic of the *Chicago Tribune*.[8]

The tubular frame would set up a wave of taller buildings, including the Twin Towers of the World Trade Center, although these were designed by a competing firm led by the Japanese-American architect Minoru Yamasaki. In 1971, the Twin Towers, supported by their tubular frames, would finally overtake the Empire State Building and claim the title as the world's tallest buildings. Exterior columns connected to the steel core through trusses in the floor. Since the columns were relatively closely spaced, the tower windows were on the smaller side—although there are some who believe this had more to do with Yamasaki's fear of heights.

Unlike the Empire State Building's columns, which were clad in terra-cotta and filled in with masonry, the columns of the Twin Towers were fire-protected with sprayed-on fire-resistant material. This led to a relatively lighter structure. The structure was strong enough to withstand the 1993 bombing, which left a crater as wide as 60 feet and several stories deep. But when terrorists on 9/11 crashed two passenger airplanes into the towers, some experts believed that this new fireproofing method may have contributed to the Twin Towers' tragic collapse. In contrast, the Empire State Building was more fireproof than a modern-day skyscraper. A B-25 bomber crashed into the 29th floor of the building in 1945 during a heavy fog. The building easily survived the impact, with many floors open for business the next Monday.

But the single tube had its own limits as well. It was close to reaching its maximum height. In addition, a building the shape of a box offered architects a limited palette of geometric shapes. New York City's Twin Towers, although simple box shapes, were more visually appealing since they were two boxes instead of one. The shape of the John Hancock tower was more intriguing because of its gradual taper. Nevertheless, these towers did not belong to our society's most geometrically sophisticated structures.

Khan offered a solution to both of these problems with the bundled tube. He referred to this concept as a grouping of straws, which are also hollow shapes (although his fellow architect Bruce Graham preferred the analogy of cigarettes, since he had them readily available for demonstration).[9] By themselves, straws are not strong. But when they are grouped, they form a stronger bundle. And if the straws are each of a different height, they would create a richer geometry than a regular tube.

The bundled tube offered a new functional improvement in structural design, allowing for greater heights, and also a more sculptural building when each of the tubes varied in heights. Khan was able to deploy the concept at the Sears Tower (now called the Willis Tower), also in Chicago. In 1973, the year the Twin Towers opened, Sears staked its claim as the world's tallest tower, bringing the title to Chi-

Willis Tower (formerly known as the Sears Tower), Chicago, SOM, 1973

cago. For his accomplishments, Khan was called the "father of tubular designs," and the "Einstein of structural engineering."

One problem with the bundled tube is that it is hard to configure
the interior, because of the large columns on the inside. The upper
floors were especially awkward, where there were fewer bundles.
Instead of the large, square-shaped bottom plans, the upper floors
have smaller plus-shaped plans with relatively shallow spaces. This
would have been fine initially, since Sears planned to occupy all the
tower's space itself—but with growth stalling at the retail company,
and with a glut of office spaces, the tower remained half empty until
the mid-1980s.

How to get structures even taller? The problem of scale creeps
in. You cannot just make a structure taller. To support a building
twice as tall, it would need to be twice as wide as well and twice
as deep. In other words, when you double a building's height, the
volume and weight increase eight times. Make a building ten times
taller, and it would get one thousand times heavier. This exponential
formula may work from a structural perspective, but it is problem-
atic for functional reasons. As the building gets deeper, a large portion of the interior space would be too far removed from the daylight
of the facade, making for a cavernous space so dim it would scare
even the shrewdest of leasing agents.

Bill Baker, Khan's successor at SOM, decided to discard the tube,
leading to a new era of "post-tube" skyscrapers. His own answer was
the "buttressed core," a throwback to the structural innovations
of the past to move toward the future. Just as the French master
masons had relied on buttresses to support their taller windows and
walls, Baker buttressed his tower, but in a new way. Three triangular
buttresses supported a hexagonal core like an upright missile, held

up by its three wings. This tripartite plan, the shape of a three-legged starfish, is more stable, similar to a tripod holding a camera.

This innovation led to the Burj Khalifa and significant gains in height. In 2010, the Burj reached 2,722 feet (830 meters), exceeding the world's previous tallest building, the Taipei 101, by a whopping 63 percent. Importantly, in contrast to the steel construction of the Willis Tower, which had for a good twenty-five years held the title of the world's tallest, the Burj was designed as a reinforced concrete building, another marker of reinforced concrete becoming the dominant skyscraper construction material. Together, the reinforced concrete with the buttressed core gave a strong backbone to the building, able to resist not only gravity but also lateral force.

While Baker had created an elegant solution to achieving larger heights, he needed a massive amount of structure inside the building to get there. In contrast, tubular designs benefited from larger unobstructed interior spaces. Although the tri-winged structure worked well for residential uses, which can have shallower floor depths, it would not quite work for an office building, which requires larger floor space and spacious ground-floor lobbies.

Soon afterward, Baker was given the chance to exceed his own record for a tower in Jeddah—in a tragic twist of fate, the same place where Fazlur Khan had died of a heart attack in 1982. With the kilometer in sight, could his buttressed core elevate our built environment to this new unprecedented height, or would Mother Nature have her way?

"IT'S ALL ABOUT EGO," admitted Prince Al Waleed bin Talal, leader of the Jeddah Tower project, although his comment revealed a bizarre economic logic. "The building itself won't yield much return. But because of its presence, the land around it will multiply in value, and we own a lot of the land."[10] He banked on an economic play borrowed from the Burj Khalifa, where the tower itself made little profit given its cost, but it helped elevate the real estate prices of the entire area around it.

The Jeddah Tower is planned to stand as the centerpiece of a brand-new $20 billion district, Jeddah Economic City. It aims to bring to Jeddah a yet-unseen level of luxury, and with it—at least such is the hope—help turn the city into a global destination, as the Burj Khalifa aspired for Dubai. Monarchs had formed a massive joint venture of real estate and construction moguls, including the Saudi Binladin Group, the primary contractor ruled by the bin Laden family—best known for the black sheep Osama bin Laden, mastermind of the 9/11 attacks in the United States.

The tower drew criticism for taking resources away from public sectors—especially given the lack of spending in infrastructure, research, and education. Critics named the tower an extreme example of "gulf futurism," a local ruling aesthetic propagating a segregated techno-consumerist landscape, not unlike the cyberpunk futures of a sci-fi dystopia.

This tall Jeddah spire had actual headwinds to deal with as well. On the ground, you might not even notice a breeze. But at the top of a 100-story building, winds could be blasting at 40 miles per hour, fast enough to break off a tree's twig. Imagine all this wind force pushing on a facade 1 km tall, or as engineers call it, a "sail area," about as large as the combined sails of the entire Spanish Armada. Whereas gravity loads determine the design of short buildings, wind loads dictate the design for tall buildings over 40 stories. The tough part of this is that, unlike gravity, which is a static load, the lateral loads of a building are purely dynamic, subject to the often unpredictable behavior of the forces of wind flow. This makes them hard to simulate and calculate. When the difference between success and failure is a 1,000-foot plummet, every measurement counts.

When a strong wind passes an obstacle, such as the edge of a building, it can start to swirl and create a strong current that pulls the structure in unexpected ways. This interplay between the building and the wind could lead to a dangerous phenomenon called vortex shedding. For instance, at lower speeds, the force of the wind on a cylindrical tower is always in the direction of the wind. But when the wind accelerates, the air begins to separate from the surface, creating two symmetrical and adjacent low-pressure eddies on the downwind side. As the wind blows even faster, the vortexes are no longer shed symmetrically but alternately, gushing from side to side behind the tower. This is an even bigger problem when the oscillating air flow happens at the building's natural frequency, the time it takes for the building to naturally vibrate back and forth. In this case, even small amounts of wind can lead to major oscillations, and could lead to collapse. This phenomenon is similar to an opera singer shattering a wine glass by singing the right note exactly at the glass's natural frequency.

The shape of the tower determines how much those currents will be. Perfectly round towers perform the worst. Square towers complicate wind flows as well, with eddies occurring around the edges. Since there is no exact science to it yet, tower designs need to be tested in wind simulations to see if a particular geometry would work.

Until the 1970s, engineers had to specify structures with thicker members than necessary, because they could not fully test their creations until they were built. Perhaps the most famous disaster was the Tacoma Narrows Bridge in Washington. Upon completion in 1940, it was the third-longest suspension bridge in the world. To save money, engineers specified a structure with shallow girders underneath the deck instead of a higher and more stable truss. This also gave the bridge an iconic and slender "steel ribbon" appearance. Unfortunately, the analogy extended to its structural performance. For its vertically moving deck in the wind, construction workers nicknamed it "Galloping Gertie." Only four months later, during a wind of 40 miles per hour, the deck oscillated in a twisting motion into ever larger waves, until it collapsed.

Today, this is a textbook example of how small periodic forces can add up to large oscillations over time, just like pushing a child's swing at the right time will make it swing higher. Most engineers believe the bridge failed as a result of its poor aerodynamics. Instead of a truss through which air could flow, large vertical steel plates on either side of the deck created alternating vortexes on the top and bottom side of the bridge, amplifying the twisting motion. The bridge then went into "torsional flutter," a self-induced harmonic vibration pattern similar to a flag vibrating in the wind. The deck's own movements became self-generating, or "self-excited," inducing greater and greater twists, and snapping the cables.

Much of our understanding of modern engineering stems from past failures. The collapse of the original Tacoma Narrows Bridge eventually led to mandatory wind-tunnel testing. Typically, sensors attached to a model take hundreds of readings of the wind pressure, and a computer analyzes this data to determine locations where the building might perform poorly. This allows the design team to make changes. The more complete the model, the more accurate the result, so entire surroundings are miniaturized, including nearby buildings, trees, and even people—all of which have the potential to complicate wind patterns. In the past, designing and building such models could take a great deal of time. Today, 3D printers can quickly churn out new models of the skyscraper, which can then be tested in the wind tunnel. "They can go through eighteen variations in a day,"[11] according to Bill Baker, the engineer of the Burj Khalifa.

The key for architects and engineers is to "confuse" the wind, and to prevent wind forces from getting organized into larger eddies. Designing structures to be aerodynamic, like China's sleek Shanghai Tower, can reduce the wind force by up to a quarter.[12] This may seem small, but for a structure as tall as the Shanghai Tower it makes for a substantially lighter structure, with the builders saving an estimated $58 million in materials costs.[13]

When done well, a wind-optimized structure can have unique aesthetic appeal. The Shanghai Tower culminates a trend of twisted

Turning Torso, Malmö, Santiago Calatrava, 2005

towers started by the Turning Torso in Malmö, designed by Santiago Calatrava. Their shapes generally help disturb the formation of alternating vortexes, while also making for good landmarks.[14] The "stair step" design of the Taipei 101, inspired by the stacked roofs of a pagoda, has softened corners to reduce vortex shedding effects. The iconic shape, reminiscent of eight takeaway noodle boxes stacked on top of each other, is the city's most photographed building. The Burj Khalifa's gradual setbacks, also inspired by the local vernacular, help reduce wind load as well. Some towers have openings piercing through the structure to reduce the wind pressure. For instance, the opening at the top of the Shanghai World Financial Center makes for an instantly recognizable shape, one likened to a bottle opener.

However, one difficulty for the Jeddah Tower was that no one had studied in detail the effect of windstorms so high above the earth. Most earthly structures lie within the planetary boundary layer. This is the portion of the earth's atmosphere where we have sensors, closest to the ground, where wind is influenced by the underlying earth surface, such as water bodies and terrain. But at the peak of the Jeddah Tower, a new layer begins, called the free troposphere. Here winds are determined by larger masses of air, instead of the earth's boundary layer, like the swirly patterns you see on the Weather Channel. Engineers had to rely on the weather balloon data from nearby Jeddah International Airport to predict the frequency of wind speeds up in the sky.[15]

Nevertheless, the balloons may underreport wind velocities, since they don't get released during storms. With these unknowns, the designers of the Jeddah Tower hearkened back to the triangular footprint from the Burj Khalifa. But to reduce wind loads, the form was more slanted. The "stayed" buttressed core arrived.

However, even with this form, the top would be swaying a few feet back and forth, potentially sickening inhabitants. How much movement is too much? Senior executives at the top of the building may be more sensitive than junior, younger staff below, since as we age, our sensitivity to motion increases. Shaking may be even more of a problem in the condominium section of the tower. You may not be too alarmed if you were to see your coffee moving in the office, since you might be too busy to notice. But you would probably react differently when you're at home, sitting at your dinner table, and suddenly your wine begins to make waves.

We now know that the maximum amount of lateral motion at the top of a skyscraper is about 1/500 of the building's height, the maximum movement a person can be expected to handle without queasiness. As buildings get older, they tend to shake more. So, what may work at first may no longer be humanly comfortable as the building ages. Buildings, too, as they get older, are more sensitive to motion, just as we are.

It took many tests to find these human upper limits. Some of these tests had very questionable ethics. Prior to the construction of the original World Trade Center in 1965, tests were carried out as far as possible from New York, in Oregon, to avoid negative publicity, in a facility labeled as the Oregon Research Institute Vision Research Center, disguised as an optometrist's office. Seventy-two unsuspecting test subjects, lured by a newspaper ad offering a free eye exam, were sent to movable rooms agitated by heavy hydraulics. Unbeknownst to the human guinea pigs, the rooms were swayed during different periods of time. Test subjects—the optometrist included—felt groggy and experienced nausea afterward, some even becoming rubber-legged. The researchers realized that humans are more susceptible to horizontal motion than they had thought.[16]

In addition to strengthening the structure, or increasing the mass of the building, engineers have some other tricks up their sleeves to reduce sway. The World Trade Center's engineers eventually decided to design the Twin Towers to sway for a maximum of 3 feet on either side. They placed more than 20,000 visco-elastic dampers between the columns and the floor trusses, to absorb the punch of the swaying motion of the building. With each shake, the dampers would move a little, and then snap back to their original shape.

Now there is an even more unique system to prevent sway. A sky-scraper is not unlike a giant tuning fork, which will start vibrating at its own resonance after a good gust. To prevent the wind from rock-ing tower tops, many skyscrapers employ a counterweight weigh-ing hundreds of tons called a "tuned mass damper." Usually a large weight on top of shock absorbers, it moves in the opposite direction of the tower's motion, like a giant pendulum. Damping takes vibra-tion out of the building and reduces it significantly.

The problem with dampers is that they are most effective at the top of the building, where they take up the most valuable real estate. Typically, they are large tanks of water or blocks of metal that are hidden in mechanical floors. Some engineers have solved this more creatively.

At Taipei 101, they turned the tuned mass damper into an attraction. They suspended a giant metal orb above the 87th floor. When wind moves the building, this orb sways into action, absorb-ing the building's kinetic energy. As its movements trail the tower's, hydraulic cylinders between the ball and the building convert the kinetic energy into heat, and stabilize the swaying structure.

For one hundred dollars you can get a tour of the Shanghai Tow-er's tuned mass damper, the first electromagnetic damper used in skyscrapers, topped with a piece of art. While you have crumpets and tea, you can witness the damper in action, under the gaze of a 25-foot-tall swinging, azure sculpture the shape of a dragon's eye.

In Hong Kong, engineers devised more practical solutions for mass dampers. Instead of a massive water tank, they placed swimming pools on upper floors that work as a liquid damper. Baffles in the water can slow down the water. By controlling the time period of the wave motion, they can dampen the building's vibrations.

At the Jeddah Tower, a tuned mass damper was planned for the top. But it may be a while before there is tuning to do. The tower's construction started in 2013. They built 270 piles to support the building, including several ones as deep as 360 feet. Above ground, construction followed in 2014. But in 2017, Prince Al Waleed bin Talal and the chairman of the contractor, the Saudi Binladin Group, were arrested on accusations of corruption. Construction stalled in 2018 with the tower about one-third complete, and people question whether it will ever be completed.

Critics likened it to the Tower of Babel, that cursed structure in the Book of Genesis. The united human race, speaking a single language, had tried to build a tower so tall it would reach heaven. God realized their arrogance had to be stopped. He checked their egos by confusing them and taking away their speech. This led to the origin of the world's many languages, and the incompleteness of the tower—likely based on the tower of the Great Ziggurat of Babylon, built in 610 BC at a height of 100 feet. In the case of the Jeddah Tower, it has more to overcome than the treacherous winds above the earth's planetary boundary layer, the limits of human comfort for sway, or the compressive strength of concrete. In the face of political turmoil, today, it stands at its reduced height, out in the desert, catching wind—perhaps constrained more by human nature than Mother Nature. Thankfully, dry desert air is not a bad place to preserve the concrete carcass, as it waits for better fortunes.

THE BUTTRESSED CORE may be just the beginning. Fazlur Kahn, in 1981, one year before his death, advanced the engineering of ultra-high buildings, determined to find "the next possible step in this evolution."[17] He applied his concept, the "telescoping superframe," to a design for the World Trade Center Chicago. It consisted of successively smaller boxes that telescoped upward, making a tapered form. Four trussed corner megacolumns tied together every twenty stories to multistory horizontal trusses stiffened each section. Portal openings throughout the structure would let the wind blow through. Unfortunately, the developer paused the project, disappearing without further contact, so it was never built.

Today, it seems that the next evolutionary step up may be a structure consisting of multiple buildings, rather than a single building with a stronger structure. A cluster of towers, connected by skybridges and sky lobbies that contain trusses the size of several stories, could together act as a megaframe. These interconnecting skybridges and sky lobbies would provide another exit option during emergencies, as well as social opportunities to interact, like sky streets between buildings.

This approach can be pictured in the Petronas Towers in Malaysia, a pair of twin towers with a skybridge in the middle, designed by architect César Pelli. At 1,483 feet, they were the world's tallest towers from their completion in 1998 up until the 2004 Taipei 101. However, the interconnecting bridge is not structurally attached to the two towers. The bridge has a "sleeve" connection, allowing it to slide in and out of the towers. Depending on the wind strength, the bridge is pulled into its sleeve in one of the two towers, to prevent it from shattering in the wind.

Only recently have true interconnected towers been built. In 2018, tech conglomerate Tencent completed its twin-skyscraper headquarters, the 825-foot-tall Seafront Towers in Shenzhen, designed by architecture firm NBBJ. Three sky lobbies offering communal space connect the two towers at different levels. Even more stable than two towers are three interconnected towers. The Golden Eagle Tiandi Towers in Nanjing, at 1,207 feet tall (368 meters), are the first trio of supertalls to together form a megaframe. A six-story skybridge at about 600 feet connects the three towers. In the future, we may see structurally connected skyscrapers rise to even bigger heights.

Shorter, gravity-defying skyscrapers are shattering other barriers as well. While Italian engineers are toiling over how to reduce the lean of Pisa and avoid collapse, elsewhere, engineers are locked in a race to vie for the most leaning building. Upon its completion in 2011, the 520-foot-tall Capital Gate in Abu Dhabi leans 18 degrees, more than four times the Leaning Tower of Pisa, winning the Guinness World Record award of "farthest manmade leaning building." It was the first building to use a "pre-cambered" core, meaning the heavy core leaned in the opposite direction of the building's incline in order to even out the forces of the building.

Several leading architects are challenging the monolithic, shaft-like appearance of tall buildings, proposing towers that "disintegrate." German architect Ole Scheeren, in his design for the MahaNakhon in Bangkok, broke up a solid tower with a spiral-shaped "erosion" of pixelated balconies and terraces. While this approach brings structural difficulties such as discontinuous columns and

Capital Gate, Abu Dhabi,
RMJM, 2011

transfer beams, aesthetically it offers a sense of individuality and
fragmentation that echoes the complexity of Bangkok city life. But
"disintegration" comes with other risks as well. In 2011, the Dutch
firm MVRDV proposed two towers in Seoul interconnected with
a pixelated "cloud." The project was abandoned after critics noted
a striking resemblance with the terrifying images of the planes
exploding into New York's Twin Towers.

Some projects are defying the linear verticality of conventional
skyscrapers. For the CCTV Headquarters building in Beijing, architects Rem Koolhaas and Ole Scheeren designed a skyscraper that
loops into itself. Two leaning towers are bent 90 degrees at the top
and bottom, making a continuous structure, nicknamed the "Big

Pants." Completed in 2012, the project required a massive building cantilever and dense cross bracing at the bends, which added to its unorthodox aesthetic.

In 2014, concerned by buildings like the CCTV Headquarters, Chinese president Xi Jinping urged architects not to "engage in weird building." Others wonder whether we really should build these extreme examples of engineering. The more we defy gravity, the more structure we need. Plus, the more materials, the more energy we need to construct these materials and the more carbon we emit into the atmosphere. The World Steel Association estimates that steel production is responsible for 6.6 percent of global greenhouse gas emissions.

MahaNakhon, Bangkok,
Ole Scheeren, 2016

Yes, these buildings may consume a vast amount of resources. But by making these moonshot efforts, we learn a lot—such as how to make lighter and stronger structures. Or how to make buildings more aerodynamic in the wind. Or how to "tune" buildings. Some of these lessons we can apply to other things.

In Toronto, designers for a proposed 35-story timber tower, a wooden exoskeleton inspired by Khan's tube, wanted to reduce the amount of structural material. They added a tuned mass damper. It would have allowed them to use only half of the amount of timber material.[18] Together, the tuned mass damper and the timber would consume less carbon than a purely concrete building would. With new and greener "plyscrapers," we may be able to apply other innovations as well.

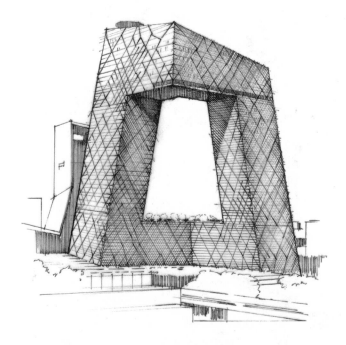

CCTV Headquarters, Beijing, OMA, 2012

Supertall towers have a structural efficiency of their own. Most engineers are excited about the design of supertall skyscrapers because they tend to be almost perfect expressions of structure. In typical projects, engineers are tasked with calculating the sizes of columns and walls when the architect has already completed most of the design. But, for a supertall, an early collaboration between the architect and the structural engineer is vital. Not least because structure is one of the key driving forces determining the shape and cost. Engineers like Fazlur Khan focused on promoting structural efficiency and minimizing material waste.

In supertalls, the geometry of the most efficient structure becomes the architecture. The more efficient shapes, as it turns out, are a throwback to history's best buildings, like the Gothic cathedrals, when master masons through trial and effort used as few bricks as possible to create the vastest spaces. Without the flying buttress gracing the Gothic cathedral, there would be no cathedral. If the geometry of the buttress is not exactly right, the building will collapse.

Even smaller buildings tend to express their structures, allowing the load-bearing skeleton to shape the building's identity. Some of these decisions based on structure have been spun into design aesthetics, like the now-iconic classic portal front, or the braces in half-timbered Tudor buildings. While initially not intended for their aesthetic appearance, today these exposed braces have become a style in their own right, with "half-timbering" plastered onto contemporary lodges and suburban homes.

Today, environmentally responsible designers need to think about architectural geometry and the shape of structure to minimize carbon emissions. In the face of climate change, each pound of material and carbon emissions may come with a tax bill attached.

There are ways of designing structures that are both materially efficient and architecturally beautiful.

In the late nineteenth century, Antoni Gaudí designed his masterpiece, the Sagrada Familia, based on material efficiency. He drew a floor plan for his domed cathedral and placed it upside down on the ceiling. He then hung metal chains from different points of the ceiling, each end connected to opposite outlines of each dome. The resulting U-shape was a "catenary curve," the curve at which the metal chain would lie at rest, with tensile forces evenly distributed between each link of the chain. When inverted, this now arch-shaped curve represents the even distribution for compressive forces. This most efficient shape for a structure where members are in compression, like a stone arch, ended up determining the domes of his cathedral.

These highly efficient shapes are also found in nature. Today, topology optimization software helps engineers find an optimal efficiency between structural strength and an economy of materials. The search for the strongest objects with the least amount of materials leads to a more organic and natural form. When you cut a bird's beak or bone, you'll find it is not solid, but has a cancellous interior consisting of a web of fine structures within its outer shell. Tens of millions of years of evolution have endowed today's birds with extremely lightweight structures. But even 20 percent of the human skeleton contains spongy bones. With a scarcity of material at its disposal, nature distributes the available material to absorb the surface tension in the most efficient way.

In short, Mother Nature has a few lessons for us humans. Billions of years of evolution and adaptation to different environments have led to some ingenious solutions, all for us to harness. For instance, after a hunting trip in the Alps in 1941, the Swiss engineer George de Mestral noticed burdock seeds clinging to his dog. Upon closer inspection, he noticed the tiny hooks that "mate" with another fabric, like animal fur, as a way to increase the seed's dispersal. Shortly afterward, based on this principle, he invented Velcro.

The field of biomimicry is dedicated to applying some of nature's design lessons to materials, products, and buildings. Material scientists, engineers, and architects draw from evolution's strongest and toughest materials and structures. Bone, teeth, nacre, bamboo, and spiderweb silk are exceptionally strong, a result of Darwinian struggle in different and harsh environments. Their qualities range from unique macroscopic forms to the nano-scale-level proteins.

With 3D printing, we can mimic these micro-scale designs of nature's finest engineering marvels. Already, shoe companies are 3D printing soles based on lattice structures inspired by insect wings and leaves. Material researchers, to make stronger body armor, find inspiration from the herringbone structure of the appendage of mantis shrimp, its weapon to smash through shells of mollusks and crabs.[19]

With new computer software and manufacturing techniques, it may become easier to make structures more material efficient. Structural members, like steel bars and concrete columns, used to be predominantly straight. Now, 3D printing can print steel in any shape, and make custom molds for more elaborate concrete forms.

With a wider palette of shapes at their disposal, engineers can devise structures that more naturally and efficiently transfer horizontal forces to the ground. Faster computer processing, artificial intelligence, and the widespread availability of simulations for wind and structural loads can help them test and predict more efficient shapes, with almost instant feedback, even on a simple desktop computer. New computational design and manufacturing methods can help engineers build even sleeker and taller buildings, overcoming the toughest of earthquakes, tornadoes, and hurricanes.

For designers of supertall skyscrapers, better aerodynamics and smarter decisions about structure may allow them to raise buildings to even greater heights. With less weight dedicated to structures, the building itself can become taller. Plus they still have some new structural forms to exploit. All of these innovations point to bigger and better structures in the future, a massive departure from our humble beginnings of stacking stones.

THREE

The Race to the Top: Elevators

Each day, a billion people around the world travel in elevators, making a total of seven billion trips.[1] That's the equivalent of moving Earth's population one ride every day. If the elevators would all suddenly stop, so would our economy. Without the elevator, there would be limited population density, and no verticality, no urban energy, no "elevator pitch." We would live in sprawling suburbs or cities limited by five-story buildings. The elevator makes modern life possible, as the great enabler of human habitation of the sky.

Unprecedented urbanization in Asia has given the skyscraper industry a big boost, leading to innovations including lighter elevator cables and faster electronic engines. China's skyscraper boom is emblematic of this urbanization drive. Within a single generation, half a billion people moved from the countryside to cities, with many people living in skyscrapers. The country has been fertile ground for what seems to be forever-faster elevators. China has the most elevators in the world, more than 7 million and counting.[2] Before 1980, it had fewer than 20 thousand.[3]

China's skyscraper construction outdoes even the Middle East.

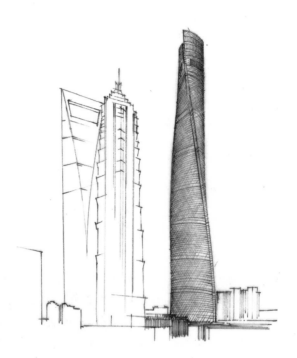

Shanghai Tower, Shanghai, Gensler, 2015

The nation may not have the distinction of the world's tallest tower, but it impressively houses five of the world's top ten. In 2019 alone, the People's Republic added 57 towers taller than 200 meters (656 feet), 45 percent of the world's total.[4]

At this "Skyscraper Olympics," the most prominent is the Shanghai Tower, the country's tallest at 2,073 feet (632 meters). Although it might not soar as high as its Middle East rival, it makes up for it by floor area—almost 20 percent more than the Burj Khalifa. And where the Burj stands isolated, the Shanghai Tower is one of three supertall spires, standing right next to the Jin Mao Tower and the Shanghai World Financial Center, in the world's first triple-adjacent supertall cluster.

The Shanghai Tower is also faster. Its high-speed elevator runs twice as fast as its contemporary, at 67 feet per second. It also runs farther, inside the world's longest continuous elevator shaft. Its whopping 1,898-feet length exceeds the Burj's shaft by 240 feet, and zips you all the way from the bottom to the top in a single shot under 55 seconds (although a South African gold mine has the world's tallest elevator that's not in a building, which is as long as 7,490 feet, or 2,283 meters, in a single descent). To get from the ground to the top would have required 3,398 stairs, and a grueling climb of 53 minutes.

The race for the fastest and safest elevator has come a long way from the slow-moving cages once used by Romans to hoist wild animals to the Colosseum floor. Today, elevators travel over 47 miles per hour, while traffic-management algorithms group riders by destination to get full and empty cabins where they are needed as efficiently as possible. With ever-lighter elevators moving ever faster across ever-thinner cables, we will soon be able to reach higher grounds even more swiftly. Future cabins may even move along frictionless magnetic rails, ditching the cable altogether, and crisscross sideways through multiple buildings.

IN 1854, DURING the World's Fair in New York, Elisha Otis conducted a groundbreaking experiment. Wearing a suit and a top hat, he stood on a platform suspended by a single rope. He hoisted himself up above the crowds. He then invited his assistant bearing an ax to do the unthinkable.

His assistant swung the ax and slashed the rope, the platform jerked, and people gasped. The victorious anticlimax followed: nothing else happened. Otis bowed, took off his top hat, and proclaimed. "All safe, gentlemen, all safe!"[5] The safety elevator was born.

World's fairs have always excited people with new wonders to change their world. The Exposition Universelle in Paris in 1889 introduced the Eiffel Tower, showcasing the strength of wrought iron. The Great Exhibition in London in 1851 brought the Crystal Palace, with the greatest plates of glass ever seen in buildings. The World's Columbian Exposition in Chicago in 1893 showcased the glowing White City, a glimmering preview of how electricity would light up cities. But in New York, few people expected that Otis's elevator safety mechanism would change the shape of cities to come. Yet the safety elevator became "the great emancipator of all horizontal surfaces above the ground floor," architect Rem Koolhaas later wrote. "Any given site in the metropolis could now be multiplied ad infinitum."[6]

While steam and hydraulic elevators already existed, they hauled goods like ore in mines, not passengers in buildings. People knew that elevators could take them up. Otis convinced them that a breaking cable wouldn't send them hurtling down.

Today it is easy to forget that elevators were once scary. I am reminded of this only occasionally when riding up in a construction elevator. The construction elevator is often located on the outside of a building's skeleton. Without the confines of the building's protection, you find yourself inside a shaky metal compartment. This rattling cage can offer quite a view as it takes you higher and higher, to the point that you may peer down on even the city's tallest buildings. Upon arrival at the top floor, a peek down the tiny gap between the elevator and the landing reveals everything down on the street thousands of feet below, with people the size of ants. This is a common sight for a skyscraper construction worker. For others, it may be a stomach-turning moment, leaving them gasping for air.

In the late nineteenth century, with the safety issue solved, the popularity of elevators took off. Otis's earlier inventions, including an automatic bread-baking oven and a bedstead-turning machine, had found little success. Ironically, his most groundbreaking invention, the elevator safety brake, came out of his frustration to move debris between floors of a New York bedstead factory. Not willing to try his luck with existing hoisting platforms, together with his sons he built his own safety elevator. They designed a clamping mechanism that gripped the guide rails. When the hoist rope would release tension in case of a breakage, a wagon spring would activate the clamping device and halt the elevator.

The success of Otis's safety mechanism was partially due to his spectacular demonstration. The year before his appearance at the World's Fair, Otis had sold only three elevators for $300 each. By 1873, the world counted 2,000 Otis elevators. Meanwhile, Otis enhanced the safety demonstration's dramatic appeal. In some performances, his assistant presented him with a saber; in others, with a pointy dagger on a velvet cushion.

His company, Otis, would go on to install elevators at the Eiffel Tower and the Empire State Building and many of the biggest buildings in the world. Today, it is still the world's largest company producing elevators and escalators.

Nevertheless, elevators had been with us for millennia before they became ubiquitous. The Roman architect Vitruvius first recorded the use of an elevator in the first century, and attributed it to the Greek inventor Archimedes around 235 BC.

This first "elevator" was a system of hoisting ropes wound around a drum, probably a block-and-tackle system of two or more pulleys with a rope between them. The more pulleys, the easier to lift the load, but the longer the rope that needs to be pulled. This principle is still used for sailing boats and operating cranes. Since these first primitive elevators were powered by human hands and oxen, easing the lift was crucial. Vitruvius named them "hoisting machines" and observed them being used in the loading and unloading of ships. He dedicated an entire chapter to the topic, which also featured another "Eureka" moment: the Archimedes screw, a pipe with a turning screw-like surface. It lifted water from low-lying areas into irrigation ditches.[7]

Elevators also belonged to the darker pages of human history. In the eleventh century, mathematician Al-Muradi in the *Book of Secrets* describes a rising "fortress demolisher," a mechanical battering ram. The ram is attached by two ropes hanging from a wooden contraption on a base of four stacked wooden scissors-like structures. As people at the base turn wheels, a rope and pulley system expands the scissors, and the ram rises up to a more suitable height to cause damage to towers. Ironically, while Otis's safety elevator enabled tall buildings, one of history's first elevators sought to destroy them.

Another forerunner to the elevator was built not to enhance passenger safety but to make a better death trap. In the Roman Colosseum, human-powered elevators brought gladiators and ferocious animals up into the arena. Up to eight men operated each elevator, pulling a series of ropes to move a wooden cage up a wooden shaft. Between 24 and 28 lifts transported up to 600 pounds each.[8] As the cage was raised to the arena level, the trap door hidden in the arena floor opened, bringing lions, wolves, and bears up from underground animal pens and passageways, the hypogeum, and into the arena. The surprise brought the crowds to roar.

Besides gladiators and beasts, the elevators also carried prisoners heading to their "damnatio ad bestias" ("condemnation to beasts"). They were tied to stakes in the arena, after which the animals were hoisted up and released. Sadly, another feat of human invention was used as a tool for humanity at its worst.

The first passenger elevators were dedicated to royal subjects. Louis XV's desire for privacy led to the first "home" elevator. He tasked Blaise-Henri Arnoult, the first theater technician to the king—who had installed the stage mechanics at the Royal Opera—with installing a "flying chair" at the Palace of Versailles. The passenger of the flying chair could herself pull the rope inside the cabinet, to raise or lower the chair with a series of pulleys and a counterweight. This counterweight lift enabled the royal mistress to get to the king's third-floor quarters from the private confines of a cabinet, avoiding a run-in with a member of the royal household. His Royal Highness had already experimented with this technology in his "flying table" at the Château de Choisy: a rising table with a meal from the kitchen below, so the king could dine in private with his guests without the intrusion of a servant.

All these elevators reached only a few stories, limited by human strength. But with the development of the steam engine during the Industrial Revolution, steam-powered elevators could reach higher grounds. In 1823, two architects built the first steam-powered "ascending room" in London, lifting up to twelve paying tourists over 120 feet to offer panoramic views over the city. The elevator was improved upon by two other architects in the 1840s, who added a belt and a counterweight for safety, as well as a rope inside the cabin that allowed the operator to stop or switch direction.

Even though many people would balk at the thought of entering an elevator, a few visionaries knew that elevator technology was on the rise—pun intended. The first elevator shaft preceded the first safety elevator. Even the year before Otis demonstrated his safety mechanism, the industrialist Peter Cooper—who had designed and built the first American locomotive—built an elevator shaft inside the Cooper Union Foundation Building in New York. Cooper knew skyscraper construction would lead to elevator innovation. Any progress in technology can drive us forward, but it will never be successful unless it is also safe. Otis eventually designed a safety elevator for the building.

It wasn't until several years after Otis's safety demonstration that elevator use would become more widespread. In 1857, three years after his demonstration, Otis installed the first safety elevator in a New York department store. It traveled at a crawling pace of under half a mile per hour. The owner spent $300 on the elevator, about $9,000 in today's money, which he didn't actually need for a five-story building. But he expected that people would want to experience the novelty of an elevator ride, then hopefully stick around the store and buy something.

Yet elevator fears persisted, not least of all because steam boiler explosions happened far too frequently. One of the most tragic incidents was a side-wheeler steamboat explosion in the Mississippi River in 1858, which sank and killed more than 250 passengers. In 1880, 170 steam boilers blew up in the United States, killing 259 people.[9] Most explosions happened in factories such as sawmills. The next most dangerous category were paper, pulp, and flouring mills, and elevators. The last thing you want is to be up at the top of an elevator shaft and suffer the explosion of the elevator's steam boiler.

The first hydraulic crane in 1846 helped mines and factories use water or oil pressure, much safer than steam, to lift heavy objects. Soon afterward, hydraulic elevators powered by water pressure raised and lowered elevators on a hydraulic piston, instead of cables. Tanks carrying thousands of gallons sitting on the roof applied pressure on the piston. Meanwhile, a second tank in the basement received the discharge from the cylinders. At the 1867 International Exposition in Paris, among the most prominent inventions stood the hydraulic elevator—in addition to the display of reinforced concrete, another enabler of skyscrapers.

The hydraulic elevator had its drawbacks. It required a pit to accommodate the piston as it pulled back, when the elevator is in the downward position. The higher the elevator was supposed to go, the deeper the pit required below the elevator. All this extra digging into the earth only to make room for a piston was not the best use of time when you want to build a skyscraper.

Electric elevators removed the burden of the pit, all while generating faster speeds. In 1880, the German engineer and inventor Werner von Siemens built the first electric elevator, laying the groundwork for his namesake corporation. In the United States, Frank Sprague developed an electric elevator as well. His elevators ran faster than hydraulic or steam elevators, and could lift heavier loads. In 1892, the Postal Telegraph Company bought elevators from Sprague and was able to run two of them at 400 feet per minute.[10] For Sprague, this was just the beginning. He claimed that his elevators could go up to 600 feet per minute, about 7 miles per hour. Though not as swift as the average office elevator speeds today at 14 miles per hour, it was still an impressive gain.

Sprague, known as the "father of electric traction," had also contributed to electric railways. Together, the train and the elevator enabled unprecedented city growth. In downtown cores, subway trains brought commuters deep into the ground, while elevators brought people through shafts to inhabit the sky. As railways helped cities grow outward, the elevator helped them grow upward.

The United States became the breeding ground for elevator

innovation, and would have the fastest elevators for much of the twentieth century. However, even with a safety device and more reliable engines, each elevator ride could have a tragic ending. Typically, elevators were in service only between 8:30 a.m. and 5 p.m., during the on-duty hours of elevator operators. Sometimes it was this very safeguard that led to disaster. The real danger lurked in the potential human error of the elevator operators. They were often not allowed to sit or speak, unless spoken to, so that they would focus on the operating mechanism.

The job required the balancing of a difficult trade-off between maximizing speed and the precision to stop. The Otis company compared the job to a "railroad engineer controlling the movement of his locomotive."[11] It became a real profession, with tens of thousands of operators employed, many of them African Americans, complete with elevator unions. The operators had to take official exams, study guides, and even attend "operator's schools" apprenticeships. Despite all this work, operators were paid relatively low wages. Ironically, due to Jim Crow–era segregation, the profession also meant that many African Americans were in charge of a machine they wouldn't be allowed to ride otherwise.

One night in 1879, a group of tourists returned to their Chicago hotel. When a woman named Mrs. Lightener arrived at her stop on the third floor, the elevator boy abandoned his spot and stepped out of the car, allowing his passengers to make a more gracious exit. Mrs. Lightener crossed the threshold, when suddenly the car began to descend. She ended up stuck halfway between the cabin and the landing.[12] The elevator operator, who was supposed to stay near the control, was later blamed by the coroner's jury for Mrs. Lightener's mangling.

Then there were cases of people trying to catch an elevator. As they jumped into the car just as the operator closed the door, they could find themselves caught halfway, depending on the operator's reflex. Due to human limitations, these incidents usually had a bad ending. Ironically, there was already a solution back in 1887, when electric elevators with automatically closing doors to the elevator shaft had been patented. But people didn't want them. They wanted operators.

Elevator music was first introduced in 1928 to make people more comfortable in the absence of elevator operators. To soothe the primal fears and anxieties that come with elevator travel—like being stuck in a small box with strangers, while being hauled upward, amid howls and squeaking. Calming music with simple melodies easy to loop, also known as Muzak, would soon find its way even outside of the cabin, to be piped into malls, department stores, cruise ships, and telephone systems to induce that mellow mood conducive to shopping.

Still, people wanted operators. Then, in 1945, a few weeks after the end of World War II, elevator operators went on strike and ground New York to a halt. A year earlier, the first elevator safety bumper had been created, able to prevent the elevator doors from closing on a passenger or another obstacle. Finally, by the mid-twentieth century, the driverless elevator would become commonplace and trusted as it is today.

The advent of elevators into everyday life left a profound imprint on the human psyche. *The Elevator* became a common name for evangelical magazines, while early elevator cabs remarkably resembled Roman Catholic confessionals, similarly confined and dark spaces.[13] The Woolworth Building, aptly dubbed the Cathedral of Commerce— in 1913 the world's tallest building—had an overall appearance and even an elaborate filigree in its elevator cab design that, according to one critic, "awakens associations with Gothic churches."[14]

The elevator became an important element in film as well, seen as a generator of coincidental encounters. Especially when people got stuck between two floors, they would find themselves in a hermetic oasis of calm within the frenetic city, an opportunity to reflect, to connect with others, and to reevaluate their lives. In *You've Got Mail* (1998), the character played by Tom Hanks gets stuck in the elevator of his apartment building with his girlfriend and two others, which leads to deep introspection and the couple's eventual separation. "An hour later, I got out of the elevator and Brinkley [his dog] and I had moved out," Hank's character said. "Suddenly everything had become clear."

With fewer elevator failures today, one of our anthropological rituals may be those moments when we don't know where to place ourselves in the elevator, leading to some sort of musical chairs of empty corners. "Two strangers will gravitate to the back corners, a third will stand by the door, at an isosceles remove," as the writer Nick Paumgarten put it in *The New Yorker*. "Until a fourth comes in, at which point passengers three and four will spread toward the front corners, making room, in the center, for a fifth, and so on, like the dots on a die."[15]

What is a mundane device today was originally seen as a promise to fundamentally alter the look of our buildings and cities. The Italian artistic group, the Futurists, were electrified by the potential of elevators and escalators (an invention born on a Coney Island pier in the late nineteenth century). Antonio Sant'Elia, in his vision for "The New City," drew a multilevel city where technological features were prominently exposed. He later wrote in the *Manifesto of Futurist Architecture* (1914): "the street will no longer lie like a doormat at ground level, but will plunge many stories down into the earth . . . linked up for necessary interconnections by metal gangways and swift-moving pavements." Buildings would look entirely different as well. "Elevators should no longer hide away like solitary worms in the bottom of stairwells," but instead "swarm up the facades like serpents of glass and iron."[16]

In reality, the economics of skyscrapers would leave most elevators hidden in the inner spaces of buildings. With a lack of daylight and views in these darker areas, they were more difficult to lease out and so became a logical place for elevators. Robbing passengers of a view, elevator design became an efficiency story, something to be minimized at all costs in order to maximize rent and not make people wait too long.

This left builders with a difficult balancing act. On one hand, they wanted to build more floors to lease out more space. But on the other hand, they then needed to build more elevators in the core, thereby losing leasable floor area on each floor—so-called efficiency.

The elevator consultant was born. Reginald Bolton wrote *Elevator Service* in 1908, in which he presented solutions for elevator design problems. For instance, how to calculate the "expected service," in terms of passengers per hour, to determine the most efficient travel pattern of elevators. Today, "elevatoring" refers to the study of how many elevators are needed and at what speed—a trade-off between time and space, space wasted for elevators, and time wasted waiting. A key metric is the five-minute handling capacity, the percentage of the building's population that can be moved in five minutes. Maximum waiting times are also determined. In an office building, this number needs to be less than 30 seconds, whereas in a residential project or a hotel, a resident is presumed to be able to wait 10 or 20 seconds more.

A few modern architects saw an opportunity to elevate the elevator. The Space Needle in Seattle, a flying saucer on a stem and, at 605

Space Needle, Seattle,
John Graham &
Company, 1962

feet tall, the tallest structure west of the Mississippi when it was built in 1962, featured scenic glass traction elevators. The saucer included a mechanical device of its own: a revolving restaurant making a full 360-degree circuit within 47 minutes—about the time of a meal.

Starting in the 1960s, the American architect and real estate developer John Portman used elevators to excite hotel guests. Departing from typical hotel design, which often featured long and dark corridors, he introduced a spacious atrium featuring glass elevators in the shape of a rocket ship. "The elevator ride was worth the whole trip," wrote architecture critic Phil Patton. "A rocket launch take off, then the passage through the building's roof to the Polaris rotating restaurant."[17] After the success of this formula in the Hyatt

Regency Atlanta, built in 1967, Portman would later fine-tune it in other hotels, including the Westin Bonaventure Hotel in Los Angeles and the Marriott Marquis in New York City.

These exceptions aside, the elevator was usually condemned to the dark interior of buildings, stuck inside thick concrete shafts. Here it would be geared for increasingly higher speeds. In 1913, The Woolworth Building was one of the first skyscrapers to have a gearless traction elevator, invented by the Otis company. A larger and more powerful motor, without a gearbox attached to it, would help elevators travel at higher speeds.

The elevator would also haul more people up in one go. The Otis company created the double-decker elevator. Two stacked cabins increased capacity, helping reduce the size of elevator cores and thus increasing the amount of rentable space. The company installed the first one in the Empire State Building in 1931, all the while running a new speed record of 1,200 feet per minute (366 meters per minute), or almost 14 miles per hour.

The original World Trade Center, completed in 1973, represented a paradigm shift with an entirely new elevator system. Otis supplied the elevators with the biggest motors, as large as 8 feet by 5 feet. But more significantly, the trip from the bottom to floor 110 was split into multiple elevators, which allowed the stacking of up to three elevators within the same shaft. After taking an "express" elevator from the ground floor lobby, you arrive at a sky lobby on floor 44 or 78, without any stops in between. From there, you can switch to a "local" elevator, not unlike a switch between metro trains in a subway. The large express elevators leading to the sky lobbies had a logic of their own, with doors on each side. People would enter on the bottom in the front, then exit on the other side, so passengers would not need to make a turn. By reducing the number of elevator shafts, the system increased the amount of usable space per floor from 62 to 75 percent.[18] It was deemed a success and was copied.

In 2005, the Otis company introduced two new inventions. Regenerative drive is a device to capture the elevator's energy during descent. By converting kinetic energy during braking into electric energy, not unlike a hybrid car, this energy is saved and released to the building's grid. Compass destination entry, or destination-dispatch elevators, helps group passengers into one cabin going to the same floor. While this system increases efficiency, the drawback is you have to make up your mind beforehand, and cannot change your destination during the ride. For some people, not having a button in the cabin allowing them to get off can be a cause of claustrophobia.

Elevator advances are steady and continuing. Remote elevator monitoring uses big data and predictive analytics to further improve elevator performance and minimize downtime. With elevator companies vying to be the fastest, most efficient, and most state-of-the-art, they are eager to find projects to demonstrate their latest innovations. Although innovation does not always mean "progress." For some, digital screens enliven elevator rides. For others, it's nothing but an annoyance and an unwanted claim on attention.

One hundred sixty years after Elisha Otis sold his first elevator, the Shanghai Tower became the battleground to claim elevator supremacy.

IT IS RARE that an elevator ride creates excitement. Yet, as people enter one of the express elevators in the Shanghai Tower, they often pull out their cameras and start recording. All eyes are on a blue television screen displaying the elevator speed, accompanied by a screen section showing the cabin's rapid ascent up the tower. The key moment everyone wants to capture is when the screen shows the cabin's 18 meters per second speed (about 40 miles per hour), although it could go up to 20.5 meters per second at its peak.

The experience is similar to riding another marker of Shanghai's modernity: the Maglev bullet train between the airport and the city. For new riders, eyes are less on outside paddy fields and construction cranes and more on an inside monitor displaying the train's high speed at 268 miles per hour (it can go up to 311 miles per hour).

At the time it was built, in 2015, the Shanghai Tower's elevator was the world's fastest, even faster than the 39 miles per hour of Disney's haunted elevator, the Twilight Zone Tower of Terror. This record is only the icing on the cake of China's meteoric rise, which is a function of rapid urbanization. Deng Xiaoping's 1978 economic experiment unleashed development on the nation that was unrivaled in the world.

Within a generation came a completely new nationwide highway system, a high-speed train network, subway systems in almost every major city, and more than a hundred million new housing units. Where less than 18 percent of people in China lived in cities in 1978, today more than 61 percent of the country's 1.4 billion people reside in cities.

Chinese cities had to grow. However, the densely populated country, about the same size as the United States and with four times as many people, could only offer these urban dwellers a more compact way of life. Unlike the United States, where most people live in single-family homes, in China, urban dwellers have little choice but to live in apartments in skyscrapers. That means a lot of elevators. In 2018 alone, China purchased about 60 percent of all the world's new elevators, close to half a million.[19]

Shanghai, the commercial capital of China, is the epicenter of this frenetic construction activity. In 1993, the city planned to create a brand-new district, Pudong, on swampland, destined to become the nation's new financial hub. To mark its center, urban planners wanted a triptych of three supertall towers.

The first, the Jin Mao Tower, designed by SOM and completed in 1999, draws from the tiered pagoda with a shape that steps back gradually. Up close, its textured facade is a feast of metal mullions and railings doubling as tracks for window washers, a modern take on the bamboo scaffolding in Asian construction projects. As soon as it was completed—at 1,380 feet, to be China's tallest building—it

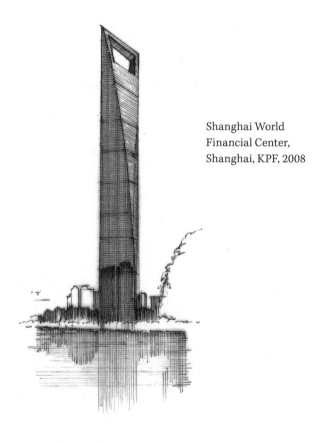

Shanghai World
Financial Center,
Shanghai, KPF, 2008

was destined to lose its title. Each subsequent tower in the triptych outdid the previous one.

The second, the Shanghai World Financial Center, by KPF, opened in 2008 at 1,614 feet. It is a tapering glass form with a large aperture on top, a circle in the original design. However, the idea of a massive circle rising above Shanghai's skyline led to a public outrage. The Chinese public associated it with a rising sun, the symbol of the Japanese flag. The designers changed the shape to a trapezoid.

The third, the Shanghai Tower, exceeded that building by 450 feet. Its structure was a challenge, let alone having elevators rise so high up in one go. Not only was the tower going to stand on the weak soil of a river basin, it is potentially shaky soil as well, given that Shanghai is located on a seismic belt—hardly the best conditions to build a supertall skyscraper. It required 980 foundation piles as deep as 282 feet into the ground, the height of Big Ben, pierced into what used to be swamp.

The structure consists of a central, nine-cell elevator core, and eight massive columns on the building's perimeter, each as wide as an average living room. They form the backbone of a shape that resembles a dynamic and twisting glass column rising up into the sky. As the tower rises, several stacked atria offer breathtaking views of Shanghai—if smog allows, that is, a tragic by-product of China's rapid development.

While the Chinese may have been uncomfortable with a Japanese shape on their skyline, they may not care—or realize—that the elevator cabin they are photographing, the elevator cables that control their safety, and the electronic engines that carry them in their ride are all made in Japan.

Japanese firms such as Hitachi and Toshiba dominate the high-speed elevator market, partly because the same high-speed elevator technology exists in bullet trains, which were pioneered in Japan.

Some observers claim that Japan is more adept at integrating new technology because of its indigenous Shinto religion, which instills spirits not just in humans but in all objects. This feature seems evident in Japan's obsession with automatic vending machines on every street corner, or the ubiquitous high-tech toilets with control panels to set the strength of the bidet splash.

When the Japanese economy suffered from the 1990s asset bubble collapse and a declining population, the most promising business opportunity lay in China. What better publicity than the ability to build the world's fastest elevator? Even with estimates for each of Shanghai Tower's 149 elevators to cost up to $3 million, it is unlikely that Mitsubishi made any money on them.[20] Instead, the company gambled on the bragging rights and future maintenance contracts.

The company faced a tall order. With longer elevator rides and higher speeds come bigger problems. One of the challenges is wind. Since supertall towers can flex up to several feet from side to side, elevator ropes can oscillate wildly, creating dangerous and uncomfortable situations for passengers. To figure out how to solve this, Mitsubishi could not rely on computer simulations alone. They needed to build a massive tower of its own.

Across the East China Sea, jutting into the skyline of Inazawa, Japan, stands a white-rendered tower, windowless, lifeless: the Shanghai Tower's surprising counterpart. In 2007, Mitsubishi built this entire tower, a massive 567 feet tall, for the sole purpose of serving as a testing ground for the Shanghai Tower's speedy elevators. The Inazawa structure was the world's tallest elevator-testing tower—

although its world record has since been overtaken by another testing tower, at 850 feet tall, built by Canny Elevator near Shanghai.

Mitsubishi designed a solution for the Shanghai Tower to pause
in times of heavy wind, plus to accommodate the shaking during a
little sway. In the time of extreme sways, sensors detect movement
and automatically stop the car at a nearby floor. This allows passengers to evacuate safely—preventing people from being trapped
in a stopped elevator.

To remove all other lateral shake, engineers created a roller system. Sensors near the cabin detect horizontal movement, and coun-
ter this by moving in the opposite direction against the vibration,
canceling out most of the shake. This elaborate system ensures that
the passengers' ride to the top is as smooth as possible—supposedly
so smooth, as the manufacturer claims, "that a coin standing on its
edge keeps [its] balance."[21]

Then there is the problem of "howling," the unnerving sound of
fast-moving winds. This first became an issue in 1963, when the Otis
elevator company installed its record-breaking elevators in the Pan
Am Building. New York requires its elevator shafts to be vented at
the top, allowing for the ventilation of smoke in case of fire. However, this leads to a strong vertical draft called the chimney effect.
When air warms inside a building, it wants to rise. The one place in
the building presenting warm air with the best opportunity to rise
continuously and fast is the elevator shaft, leading to howls.

In addition, the chimney effect can also lead to a partial vacuum,
which presents engineers with an additional problem. Not only did
the Pan Am have howling elevators, it also had rattling elevator doors
and creaking interior walls. But it could be worse. At the World Trade
Center, the vacuum was so strong it required two men to open the
doors on the ground level. Once it opened, a massive blast of air made

it virtually impossible to close. The solution was an air lock system at the elevator cabins, to avoid currents from entering the shaft.

All these innovations and more came together at the Shanghai Tower. Mitsubishi Electric streamlined the top of the cabins with fairings, curved shells that improve aerodynamics, typical of racing motorcycles, to reduce wind noise and howling. Engineers also devised a solution to the problem of people's ears popping, a result of air pressure at the top being about ten percent lower than at ground level. They installed a system to control air pressure, giving ears more time to adjust.

As great as all of this may be, what about passenger safety? It has been a long time since Otis's wagon spring–activated clamping mechanism. In Hollywood, the rope carrying elevators is never far from snapping, sending the car down the ground at breakneck speed. Given a hundred and fifty years of elevator safety advancements, odds of a fall are slim.

The typical elevator is carried by four to eight ropes, each of these, as often mandated by local elevator code, strong enough to support the weight of the car and the counterweight by itself. In the worst case, even if all ropes were to break, the elevator would be stopped by safeties, an inbuilt braking system.

As a result, elevators are the safest form of motorized transport. In the United States, about thirty deaths occur each year, most of which are, unfortunately, of construction workers.[22] The few elevator deaths are rarely due to a snapping cable, but mostly from someone unwittingly stepping into an elevator shaft without the cabin present, or from elevator surfing—the bizarre hobby of riding on top of elevators, which may even involve a jump between two moving elevators. Even when a B-25 bomber flew into the Empire State

Building in 1945 and sheared an elevator's cables, the elevator operator survived a fall of 75 stories—a Guinness World Record for the
longest survived elevator fall.

But the Shanghai Tower goes much higher than this, at 126 stories, and runs about twenty times faster than a regular elevator.
To allow for an emergency stop at higher speeds, engineers provided double safety gears with fine ceramic components that can
withstand friction heat as high as 1,000 degrees Celsius. Even if
everything else fails, a telescopic buffer at the bottom of the shaft
functions like a massive cushion.

With these problems solved, Mitsubishi managed to claim the
record for the fastest elevator. Since the Pan Am Building, the company has broken the world record several times, first in 1978 with
elevators in Tokyo's Sunshine 60 building reaching 22 miles per
hour, and then in 1993 with those in the Yokohama Landmark Tower
reaching 28 miles per hour. In 2004, Toshiba overtook the record
with the Taipei 101 at nearly 38 miles per hour. So, in 2016, with
the Shanghai Tower traveling up to 46 miles per hour, Mitsubishi
reclaimed the record . . . but only for a year. In 2017, Hitachi overtook Mitsubishi's record at Guangzhou CTF Finance Center, with
an elevator reaching 47 miles per hour. "I get tears in my eyes," said
a Hitachi sales supervisor, proud of the CTF record.[23] The elevator
passengers only experience a smooth ride to the top.

HOW MUCH FASTER can elevators go? The ultimate limit for elevator speed may be human. Some believe the limit lies around 54 miles
per hour, when people would not have enough time to adjust to the
air pressure when they got out at the top.[24] The downward limit is

much lower, at 22 miles per hour—go any faster and the body may think it's falling. To get beyond this speed limit, the entire building would have to be pressurized, like an airplane.

The real limit for elevators is not so much their speed but their acceleration. If they traveled at a constant speed, like the 190 miles per hour in a high-speed train, you would not feel any movement. But people are sensitive to acceleration, the force that presses you down on the floor as the elevator starts rising, and deceleration, the float-ing feeling when it's braking. As you accelerate downward, you could feel almost weightless, like an astronaut in space. Experts think the limit of human comfort is at about a five-foot increase every second.

Of course, what is the point of such a fast elevator for anything other than a one-way ride to the observation deck? For any other use in the building, like getting people to their floors in between the bottom and the top, such a fast speed may be worthless when you add in the time necessary for stopping and loading and unloading passengers. It's like driving a Ferrari downtown, with traffic lights on every block.

For some, the real future of elevator advancement is not in speed, but in the length of a continuous ride. The limitations for shattering this barrier lie not in human endurance but rather in cable strength. Here, the weight of the elevator cable is the limiting factor. The taller a building, the longer the elevator cable, and the more of its own weight that cable needs to carry.

The ghosts of cables past include a hemp rope, proposed by Otis and used in the ancient elevators of Archimedes. Metal wire rope became dominant in the 1870s, where several strands of wire were twisted in a helix shape. But the Shanghai Tower's rope, the so-called sfleX-rope, is thicker and denser than your typical wire, combin-ing steel and resin, which allows more load while only marginally increasing the cable's weight. Nevertheless, without a better rope, engineers think the conventional hoist elevator has reached its limit.

Now there may be new hope. The International Academy of Astronautics has proposed a carbon nanotube "rope" as part of their 100,000-km-long "space elevator." This tether, anchored to the Earth's surface, would project far into outer space. Climbers would move up the cable without rockets, reducing the cost to launch satellites. In 2019, the International Academy of Astronautics published a report stating that such an elevator may be nearer than we think, thanks to a potential new manufacturing process of crystal graphene, a material about a hundred times stronger than the strongest steel.

Back on earth, the solution for hoist-less elevators may lie in pneumatic vacuum elevators, first used in 2000. Turbines at the top draw air from the tube and suck the cab upward. Other than a vacuum tube, such an elevator requires no pit, hoist way, machine room, or cable. It's a bit like an old-time pneumatic tube mail system used to propel people instead of letters. Thus far, these elevators only travel up a few stories at best. With the development of the ultra-fast Hyperloop train, which travels through a low-pressure sealed tube, perhaps one day we will see a Hyperloop elevator.

Maybe there is an end to this elevator madness after all. Several companies have decided their time and ingenuity is better spent on creating "green" elevators. Elevators are one of the most energy-efficient means of transportation, since you can fit in more people when you go vertically, instead of horizontally. (When you accelerate vertically, you get compressed into the floor. When you accelerate horizontally, you take up more space, with people spreading their stance.) Still, elevators typically consume 2–10 percent of a building's total energy use, and during peak hours this number can be as high as 40 percent.[25]

Regenerative drives to capture braking energy are becoming more commonplace. They can reduce elevator energy consumption by up to 75 percent compared to conventional elevators, according to the Otis company. In New York City, the 2020 energy code forces all new tall elevators to have a regenerative drive. In light of climate change, speed seems a wasteful exercise.

Maybe those in the race for taller and faster elevators need a little out-of-the-box thinking. The real limit for elevators is to move beyond their up and down motion. The future of the elevator is sideways. The sideways elevator is considered the "holy grail" of the elevator industry. The German company Thyssenkrupp has developed one called MULTI. It had been nicknamed as the "Wonkavator" after Willy Wonka's sideways lift, a radical departure from "just an ordinary up-and-down lift. . . . This lift can go sideways and longways and slantways and any other way you can think of! . . . You simply press the button . . . and zing! . . . you're off!"[26]

Now they have already built a working model in a test tower in Rottweil, Germany (yes, home of the dog breed). Departing from the rope elevator, the MULTI is driven by magnets, like the Maglev train, which uses electromagnets to levitate the train slightly above the track and propel it forward. Relieved from the burden of the rope, this magnetic elevator might go higher than ever.

And, just like Willy Wonka's creation, a cabin on magnetic tracks can go in other directions besides up and down, such as diagonally or sideways. A rotating "exchanger" can move the MULTI from one track to another, not unlike a train. This also allows multiple cabins to share tracks, instead of the sole rope and shafts that are dedicated to each traditional hoisting elevator. With multiple MULTI elevators in a single shaft for new buildings, this will allow for a more economical use of space, and will reduce wait times.

This is just the beginning. Imagine a society with a sideways elevator. Instead of skyscrapers going up vertically straight, constrained around an elevator core, they could be shaped like an octagon or an X, with elevators rising up diagonally. Virtually every letter in the alphabet could become an architectural shape.

Or skyscrapers could seamlessly integrate with underground metro stations, with magnetic cabins going from subway tunnels directly into buildings, dropping you off at your office floor. Buildings could be interlinked with sideways elevators, creating a network of skybridges and a more efficient way to travel. People with disabilities could more easily move around buildings or entire districts.

It could be possible for the cabins to actually be your office, and you can ride them anywhere you want—conceptually not too different from the early elevators, considered "movable rooms" with cushioned benches and carpeting. This magnetically levitated future may sound like an episode from *The Jetsons*, but a sliver of this may be here faster than we think. Probably first on university campuses or in large hospitals, where there may be bigger budgets and greater control to do this than in a city.

Before this happens, an army of MULTI consultants will need to overcome hurdles in the elevator code, as the company found in the North American applications. It took officials some time to figure out whether this new device should follow codes for trains or elevators.[27] The problem with innovation is once again human. In this case, not what our bodies can stand but what our bureaucracies can tolerate in the name of safety.

The elevator has come a long way since Archimedes wrote about the block and tackle. But it may be a while before we find ourselves pressing a "sideways" elevator button, and doing an elevator pitch in our magnetically levitated ride between two buildings.

FOUR

The Cooling Effect: Air-Conditioning

Frank Lloyd Wright drew many detailed blueprints of his Mile High tower, including one at 1/16-inch scale—being itself as tall as a two-story building. Nevertheless, his project was an abstract idea in a technological sense. Without a state-of-the-art cooling system, a glass tower would effectively become a giant microwave.

Supertalls house myraid activities that together generate a lot of heat. In addition, as skyscrapers rise above other buildings, they are fully exposed to the beaming sun. Since many building owners opt for glass facades to get more views and that crystalline aesthetic, ultraviolet rays gush in.

Opening a window can be impractical and can change airflow, causing doors to slam shut or be sucked open. For these reasons, skyscrapers need another machine of modern convenience, besides the elevator. They require the air to be artificially cooled in all seasons. They need powerful air-conditioning.

Air-conditioning systems were around in Frank Lloyd Wright's day—indeed, he specified air-conditioning ducts in the floor slabs tapering from the core of his Mile High skyscraper. But a building so high cannot use the average means of climate control. At the mercy of gravity's forces, pumping cooling liquid all the way to the top in a single shot would amount to two thousand pounds per square inch (psi). This is about as much pressure as the water-coolant system inside a nuclear reactor.

However, more recent technological innovations make supertall cooling possible. Cooling large buildings has come a long way from the brute-force efforts of air-conditioning pioneers relying on shipping ice and blowing air past ice blocks. Mankind, in the twentieth century, was able to create its own climate—until it backfired.

"Climate" is derived from the ancient Greek "clima-ata," defined as the "slope of the earth from equator to pole," a function of the earth's sphere and its relation to the sun.[1] Although the modern view of the spherical Earth did not emerge until the sixteenth century, the Greek philosopher Pythagoras had already understood this in the sixth century BC. He realized that closer to the equator, at a closer distance to the sun, life becomes unbearable. Parmenides, his disciple, demarcated five zones on this sphere. He noted that the central hot zone was uninhabitable because of the sun's rays. Some areas, Mother Nature told us, were not meant for us to inhabit.

With modern air-conditioning, the convention changed. Previously undesirable areas became comfortable and highly desirable, from Singapore to Dubai, Phoenix to Las Vegas. In some cities, you can spend your entire day moving from your air-conditioned home to your air-conditioned vehicle to your air-conditioned office, without being exposed to the actual weather. We even air-conditioned spaces as vast as entire football stadiums and cool enough to enable indoor ski slopes in the desert.

We started building taller and taller glass towers that bask in the sun yet are kept to a constant interior cool. We thought our race to the top could not be stopped. Then the climate changed.

Our desire for cool air comes at an enormous environmental cost. The building and construction sector is responsible for 39 percent of energy-related carbon dioxide emission, according to the UN Environment Programme and the International Energy Agency.[2] Most of the carbon emissions related to the building and construction sector stem from the energy to cool, heat, and light our buildings. In the U.S., the electricity use for cooling commercial and residential buildings alone totals 10 percent of the entire country's electricity consumption.[3]

Global cooling is causing global warming. We then need to cool even more. So the vicious cycle spirals further down.

Yet, if we were suddenly to pull the air conditioner's plug, our modern world would grind to a halt. Without the air conditioner, there would be no shopping malls or movie theaters, their deep floor plates trapping heat like an oven roaster. Entire cities would become unbearable. There would be no internet were it not for data centers kept artificially cold. Even a glass skyscraper in a temperate climate cannot survive without being attached to a life-support machine called air-conditioning.

A better approach would be to rethink the design of our buildings. If we don't, we may permanently lose our cool.

IN 1902, the young American engineer Willis Carrier was tasked with straightening out a production snag at the Sackett & Wilhelms Lithograph and Printing Company in Brooklyn, New York. Over the past two summers, high humidity had plagued the printing company, leading to misprints. These included misregisters of multicolor prints in some of its most iconic publications, such as the satirical magazine *Judge* (the paper later known for hiring the cartoonist who was to take up the pen name Dr. Seuss).

Cellulose fiber, paper's major substance, shrinks and expands as the air's humidity changes. At the printing company, as the humidity fluctuated, each color register of the ink misaligned with the paper's expansions and contractions. Carrier had made a name for himself engineering drying and heating systems. He knew that a constant 80 degrees Fahrenheit with a relative humidity of 55 percent would be ideal for printing. When he calculated the average humidity for New York City in summers, he realized he would need to remove about 60 gallons of water from the air, every single hour.[4]

To solve this problem, Carrier first created a system with two sets of coils. One set of coils contained a refrigerant, a liquid that evaporates as it absorbs heat—like water does, but at a lower temperature. After absorbing the heat from the air inside the room through the coils, the refrigerant, now a gas, is pumped outside the building. There it dumps the heat onto another coil. As that cools, the refrigerant gas becomes a liquid again.

His system was a breakthrough. "Taken together, their cooling effect totaled 54 tons, the equivalent of melting 180,000 pounds of ice in a 24-hour day ... a milestone in man's control of his indoor climate," remembered Carrier engineer Margaret Ingels.[5]

In this cooling process, the indoor air is dehumidified as well. Moisture in the air condenses on the cool coils, since cool air cannot hold as much humidity as warm air can. The condensate water was then drained, and the air's humidity content dropped.

While this system allowed for cooling and some form of humidity control, for Willis it was not good enough. Willis recalls how he came to a better invention, as he stood on a foggy train platform and stared at the mist. "Here is air approximately 100 percent saturated with moisture. The temperature is low so, even though saturated, there is not much actual moisture.... If I can saturate air and control its temperature at saturation, I can get air with any amount of moisture I want in it. I can do it, too, by drawing the air through a fine spray of water to create actual fog."[6]

His idea itself was counterintuitive: to dehumidify air with water. By spraying air with a fog of cold water, he could create dry air. Cold mist would lower the temperature of the air past the dew point, forcing the moisture in the air to condense and become liquid water, which he could drain. If he wanted to increase the humidity instead, he would need to create a mist with warm water, so the air would stay above the dew point.

Carrier went on to create a spray-type air-conditioning system that could both humidify and dehumidify the air. His machine would first chill the water before spraying it into a chamber. This mist of many tiny droplets would cool and dehumidify the air. After the misting of the air, it was blown through another chamber where baffles separated the water droplets from the air.

Initially, in a paper in 1911, Carrier defined air-conditioning as "artificial regulation of atmospheric moisture" with productivity benefits.[7] "In many other industries, such as lithographing, the manufacture of candy, bread, high explosives and photographic films, and the drying and preparing of delicate hygroscopic materials, such as macaroni and tobacco, the question of humidity is equally important."

Architecture critic Reyner Banham later described Carrier as "content to solve problems as they were put to him—often with startling ingenuity and depth of technical or intellectual resource—that one may doubt whether he had any general means of conception of the art he was founding until long after he had fathered it."[8]

More so than moisture control of macaroni, his machine would go on to change how we lived. Carrier only later realized that his invention provided benefits beyond regulating humidity alone. In 1915, he formed the Carrier Engineering Corporation. Today, Carrier is among the largest manufacturers of air-conditioning equipment in the world.

Carrier's cooling system was much better than previous, primitive cooling devices. More than four thousand years ago, ancient Egyptians hung wet cloth and moisturized reeds in doorways and windows, so when the wind blew across them, the air cooled through evaporative cooling, since water absorbs heat in order to evaporate. The Romans circulated cool water from aqueducts through the walls of wealthy citizens' homes . The notoriously decadadent Emperor Elagabalus forced Roman slaves to haul "a mountain of snow" for his sweltering villa from distant mountaintops. In 775, Caliph al-Mahdi of Baghdad specified hollow walls for his summer house, to be packed with snow carried by a train of camels.

By the late nineteenth century, more promising systems used condensers to liquefy refrigerants, such as liquid ammonia. As the liquid evaporated, it absorbed heat, and the temperature dropped.

Even Benjamin Franklin, before he helped draft the Declaration of Independence and the Constitution, was something of an air-conditioning pioneer. Together with a Cambridge University chemistry professor, he conducted an experiment in which he evaporated highly volatile liquids, such as ether, and brought the ball of thermometers past the freezing point. He once succeeded in bringing the thermometer's temperature down from 64 degrees Fahrenheit to 6 degrees Fahrenheit (-14°C) and noticed a coat of ice forming on the bulb. "From this experiment one may see the possibility of freezing a man to death on a warm summer's day," he concluded in 1758.[9]

With refrigeration, steam-driven condensers and compressors could produce ice, avoiding the harvesting of ice from frozen lakes and carrying it on freight trains. Air-conditioning systems would force air over this artificially generated ice, pumping the cooled air into spaces. New York's Madison Square Theatre used tons of ice per performance. In downtown Denver a two-mile-long underground pipe system offered liquid ammonia to local restaurants, hotels, and saloons, producing artificial ice, refrigerating beer vaults, and cooling bottles of champagne.[10]

While that method was better than nothing in the summer heat, ice was impractical and allowed little control over the air's humidity. Carrier avoided the need for buildings to rely on giant ice machines. However, his system required coolants that were either highly toxic, like ammonia or methyl chloride, or highly flammable, like propane. Sometimes the refrigerant leaked—for instance, from machine failure, vibrations, or servicing procedures. This would lead to immediate evacuation. If the refrigerant was ammonia or sulfur dioxide, people would suffer from vomiting and burning eyes. Methyl chloride even caused deaths.

Eventually, General Motors and its division Frigidaire found a nonflammable and nontoxic refrigerant. In 1930, GM partnered with DuPont to produce CFC Freons, a trademark name that now refers to chlorofluorocarbon (CFC), hydrofluorocarbon (HFC), and hydrochlorofluorocarbon (HCFC)—the latter of which is mostly used in home cooling, since it is easiest to compress. Air-conditioning could now be safe.

Unfortunately, history repeated itself. Half a century later, governments banned Freons, when scientists found they caused ozone depletion. Companies invented new refrigerants to not damage the ozone layer, like R-410A, also known as Puron. This is simply a lesser evil, with high global warming potential instead.

Nevertheless, with Carrier's system of coils and dew-point control, together with DuPont's refrigerants—first used by Carrier in the 1932 home air-conditioning unit called the "Atmospheric Cabinet"— the vapor-compression cycle of the modern air conditioner became possible. This system works on the phase change of a refrigerant between liquid and gas to move heat.

Where Carrier's initial air-conditioning machines may have worked for small spaces, the cool air would not reach far enough for large spaces. Carrier needed a network of ducts to carry the cool air deep into the interior and then return the hot air back to the exterior.

The first ducted air conditioners were used in movie cinemas, with great success. Many of those cinemas had been closed in summers before, since they became too hot. The air-conditioning became a reason on its own to visit the cinema, just to find a cool spot in summer (a tradition that continues today).

However, when Carrier wanted to install air-conditioning in his own Pennsylvania offices in 1928, he realized a problem. The bigger the office space, the larger the size of the duct. This would make it difficult to ventilate an entire skyscraper. "The very reason for the skyscraper was to provide more floor space on a comparatively small plot of expensive ground," wrote Ingels. "Hence, ceiling heights and floor areas were of much concern to owners. . . . They were chary of sacrificing cube-footage to bulky ducts."[11]

Carrier engineered his way out of the problem by circulating high-velocity air instead, so he could bring more cool air in smaller ducts. He commercialized it in the "Conduit Weathermaster System," invented in 1939. The system was ideal for large buildings and skyscrapers. His biggest client was the Pentagon, opened in 1941 as the world's largest office building. Carrier's system would change the design of buildings. In 1932 a Carrier publicist looked forward to the day to "enable the prescient architect to go about his designing unfettered by the erstwhile necessity of 'ventilating shafts,' 'light wells,' 'outside exposures,' and such considerations."[12]

Sadly, all of these design considerations are hallmarks of sustainable architecture. With modern air-conditioning, gone were office footprints with "H" or "T" shapes to let in light. Instead, the modern office building became a bulk square block. Each floor plate would be as deep as possible, and windows would no longer open.

In 1952, Carrier sponsored a house with a Carrier air-conditioning system, promoting a separation between architecture and climate. "This house started a revolution. It need not depend on natural ventilation. Ells and wings wouldn't be necessary. Only a few windows need have a movable sash. The bathrooms needn't require a window. Windows, doors, and even rooms themselves could be placed to suit the convenience of the owner, not to catch a breeze."[13]

With air-conditioning, employees didn't need to sit as close to a window. This allowed for larger floor plates. While this might improve collaboration between people, it also increased energy inefficiency. Large floor plates could no longer be cross-ventilated, thus making them more dependent on air-conditioned ducts. The deeper the space got, the less natural light could penetrate the building's interior. This required more electrical lighting, which made the space hotter, and therefore required even more cooling.

Air-conditioning made buildings more obese, further increasing their dependency on air-conditioning. It led to even larger air-conditioning systems, an entire new industry, and a profession. Architects designing a building would not have to worry about indoor climate. HVAC consultants (heating, ventilation, and air-conditioning) would ensure cooling, specifying the size of ducts and the mechanical equipment.

Banham lamented this. "In a world more humanely disposed . . . [it] would have been apparent long ago that the art and business of creating buildings is not divisible into two intellectually separate parts—*structures*, on the one hand, and on the other *mechanical services*. Even if industrial habit and contract law appear to impose such a division, it remains false."[14]

Where traditional architecture encompassed environmental concerns holistically in the building's structure, modern architecture became an exercise of sleek aesthetics and functional openness. This was professed by the design school of the Bauhaus, which featured one of the first curtain walls, a nonstructural all-glass facade hanging from the floor slabs. It had large open floor plates to encourage innovation. Yet all this transparency came at a cost.

The school had emerged in a time with increased access to coal.[15] Because of its large single-pane glass windows, the building had a poor thermal system, with lots of heat lost. Radiators were directly exposed to glazing, to provide a barrier from the cooler temperatures coming from the windows.

The aesthetic later became formalized in a movement called International Style modernism, which preached a universal architectural style, often regardless of local climatic context. Large, unobstructed glass windows may work well in Dessau, but less so in Dubai.

Globalization promoted architectural standardization, air-conditioning included. The American Society of Heating, Refrigerating and Air-Conditioning Engineers developed ASHRAE 55, a standard for air-conditioning to determine the ideal temperature for human occupation. This was derived from a formula from 1960s

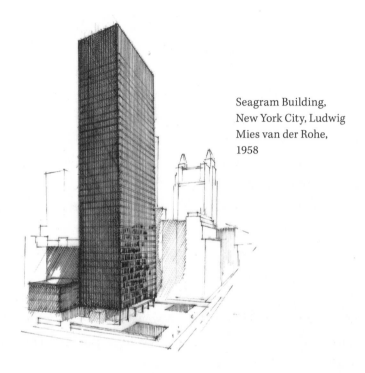

Seagram Building, New York City, Ludwig Mies van der Rohe, 1958

research, which was based on the metabolic rate of an average male wearing a suit. As a result, air-conditioning systems worldwide were set at temperatures that left many women shivering in the office.[16] The patriarchy won again.

Meanwhile, air-conditioning's dominance grew with the rise of taller glass boxes. Architect Ludwig Mies van der Rohe, the last director of the Bauhaus, designed the Seagram Building in New York City, completed in 1958. The 515-foot building was a minimalist black-glass box—a pure expression of his architectural mantra, "Less is more." He designed this air-conditioned behemoth despite already knowing the trouble with glass boxes on a smaller scale, having been sued over his Farnsworth house design by the owner, Dr. Edith Farnsworth. The house was over budget, and as it baked in the Illinois sun, it racked up cooling bills.

Nevertheless, the Seagram Building quickly became a standard skyscraper look, especially for corporate headquarters. It was so influential that, as the writer Tom Wolfe pointed out, soon New York's Avenue of the Americas would see "row after Mies van der row of glass boxes."[17] For its chilling lack of detail and austerity, Wolfe named the glass-walled avenue "Rue de Regret."

The glass aesthetic also appealed to institutions, including the United Nations. Modernist architect Le Corbusier codesigned the UN Secretariat Building in New York, a 505-foot-tall glass colossus. He argued for sun-shading screens to avoid too much heat from the sun. "My strong belief is that it is senseless to build in New York City, where the climate is terrible in summer, large areas of glass that aren't equipped with brise-soleils," he said. "I say this is dangerous, very seriously dangerous."[18] But it was not to be. Without the sunbreakers, the building became a heat trap, costing up to $10 million to heat and cool each year.[19]

Ironically, the minimalist aesthetic did not go well with the maximalism of air-conditioning ducts, electrical wires, and mechanical rooms. They were hidden at all costs. False ceilings concealed ducts. "Shadow boxes"—opaque spandrel glass behind clear glass—created the illusion of depth in the facade, hiding ducts, wires, and entire mechanical floors from the facade. Architectural louvers disguised ventilation outlets. Architecture became an exercise in making mechanical systems invisible.

Few architects of the time dared to express these mechanical systems in the facade, as was done in the Centre Pompidou in Paris, a museum dedicated to contemporary art. Architects Richard Rogers and Renzo Piano deliberately exposed the ducts on the exterior of the building and painted them in bright colors. They managed to turn building services into an art form, creating a lasting icon of "high-tech architecture."

Almost all glass buildings were interested in showing off their forms while hiding their essential mechanical system. Their minimalist aesthetic facades belied what was seen as an ugly necessity, a zigzagging maze of ducts and machinery.

In the spirit of Carrier, mechanical engineers would solve the unique air-conditioning challenges taller buildings brought. A regular high-rise might rely on a single system with a chilling plant, cooling tower, and pumps to move the chilled water through pipes up the building. A building that's breaking the supertall size barrier puts an increased gravitational pressure on the vertical pipes. These buildings need interstitial mechanical floors, to avoid dealing with massive pressures that would lead to prohibitively expensive powerful pumps.

Most supertall skyscrapers have entire floors dedicated to mechanical equipment every 20 to 30 stories. They contain large pumps, chiller plants, and electrical transformers to distribute the chilled water around the building. Each pump on the intermediate floor acts as a relay for the one above. The Burj Khalifa features seven double-level mechanical floors in total, spaced out every 30 stories, and moving water in 21-mile-long pipes to supply the air-conditioning alone.[20]

Getting the equipment up into the tower is a challenge in and of its own. Some of the chillers alone are the size of a small truck. Tower crane operators are tasked with hoisting these massive plants high up the tower. It can be tricky to lift the chiller all the way up and in between the floor slabs, when a single gust of wind could send it crashing into the building.

Then there is the challenge of how to create air-conditioning for large skyscrapers with different functional occupancies—for example, a section of the tower dedicated to a hotel and another to an office building, each with their own mechanical requirements. To solve this issue, engineers parcelized the tower into independent, vertically stacked zones, with each use served by its own dedicated chiller and boiler plants.

Massive all-glass towers built in sunny environments will never be exemplars of green building, yet engineers have nevertheless toiled to increase energy efficiency. The large investment of such a tall building warrants the exploration of energy-saving strategies, such as double-skin glass facades. For instance, the Shanghai Tower's double curtain wall encapsulates atria, which provide a buffer against extreme temperatures and contribute to the building's overall energy reduction of 21 percent.[21]

Mechanical engineers also exploited the unique properties of a supertall building's extreme height. Since the outdoor temperature on a kilometer-high building can be up to 12 degrees Fahrenheit lower at the top (known as the lapse rate), each floor has a different design condition. Just as Pythagoras realized the world can be divided into zones, engineers use the same concept with the climate of buildings. Separate technical zones cater to the outdoor air on each level having its own temperature. This helps reduce energy. At the Jeddah Tower, for instance, air-cooling energy is planned to be about 40 percent lower at the top because the air intake is cooler.[22]

Engineers even figured out how to use the cooler evenings to their advantage. During off-peak hours, the Burj uses energy to create a vast and slushy reservoir of ice slurry, a mixture of small ice particles and chilled liquid. During peak hours in the daytime, this thermal energy is used to reduce air-conditioning power consumption. The system reduces the size of the chiller network by a third. By delivering water at lower temperatures than regular chillers, it allows for chillers with reduced capacity and smaller pipes and pumps.

Engineers even created ways to recycle water condensation. The Persian Gulf suffers from high humidity, at times reaching 98 percent. The air-conditioning plants in the Burj Khalifa generate roughly 191,000 liters of condensed water per day, more than 1,200 bathtubs.[23] This can cause damage to the electrical system and can discolor windows. A separate system was implemented to collect all the condensation water. It carries the water through pipes to a holding tank at the base of the tower, where it is used to irrigate the Burj Khalifa park.

All of this is a drop in the bucket. Using the condensation from the world's tallest glass tower's air-conditioning system to irrigate the desert only gets us so far. We need a new design paradigm for skyscrapers that goes beyond crystal palaces.

IN CONTRAST TO THE EVOLUTION of air-conditioned behemoths, a small strand of modern architecture advocated for buildings easier on the environment. About two decades after the invention of air-conditioning, in Germany between the wars, architects began designing energy-efficient buildings. Their motivation stemmed from a concern not with cooling but with heating.

Germany had lost access to many coal mines when Allied forces occupied the Ruhr. This meant that architects could not rely on mechanical heating. Instead, they shaped buildings to harness the energy of the sun. In Frankfurt, an entire suburb was created, Frankfurt am Main, by noted architect Ernst May, in such a way that each apartment would receive ample sunlight. The architects promoted heliotropic design principles, arguing for *"licht, luft und sonnenschein*" (light, air, and sunshine)—and probably to a fault, ignoring other principles like site design.[24]

The buildings were oriented toward the south. They were spaced far enough apart from each other that, even when the sun was at its lowest point in winter, during winter solstice, the lower apartments would not be cast into shadow by the adjacent building. There would be enough sunlight to naturally warm the building.

With this functional, if not human-scale, urban design, the solar house movement began. Several architects around the world experimented with solar-oriented architecture in the 1940s. They created buildings that turned to nature for heating and cooling. Instead of mechanical systems to heat or cool their buildings, they used smart design to dissipate or harness heat from the sun. This method is called "passive solar" design because it doesn't require any energy-powered devices. Many vernacular buildings demonstrate these principles; however, they were largely forgotten in the age of abundant energy. For instance, builders of vernacular tropical architecture built their homes with open porches and high ceilings to create comfort through cross ventilation. The ancient mud skyscraper city of Shibam in Yemen, built in the sixteenth century and known as "Manhattan of the Desert," has a densely packed design that keeps buildings in the shade, while thick earthen walls absorb intense daytime heat and release it in the cooler night.

Passive solar architects benefit from their understanding of local climate and the movements of the sun. For instance, in the Northern Hemisphere, the sun gets to its highest point at noon in the south sky. In temperate climates, if we were to orient a building with windows opening to the south, the sun's rays would keep the room and its occupants warm.

While we may want this during the cool winter, in the hot summer we wouldn't want to add any more heat. Since the Earth tilts on its axis by about 23.5 degrees as it travels around the sun each year, the sun's altitude above the horizon at noon is 47 degrees higher in the summer than in the winter. A well-designed shading device above a southern window can do the trick. It should be long enough to block the high summer sun but short enough to allow the lower sun rays to heat the building in winter, when the extra heat would be welcomed.

Frank Lloyd Wright used passive solar principles in his Jacobs House in 1944, otherwise known as the "Solar Hemicycle." He created an arc-shaped house with wide overhangs on the south roof that reduce solar gain in summer. Modern calculations show that the house design still manages to achieve a 53 percent energy saving during the Wisconsin winter, when temperatures typically reach a chilling 7 degrees Fahrenheit.[25]

Some architects devised more technical solutions to the problem. In the United States, the first passive solar home, Solar I, was completed by MIT faculty on campus in 1939. It looked like a typical house, except for the roof. It was covered in glass-covered panels containing copper tubes with flowing water, bringing solar energy to a massive 17,400-gallon tank underneath, from where it would warm the home.

Interest in these solar homes grew in the 1950s, when it became clear that our oil supply was not infinite but would peak and then dwindle. Victor and Aladar Olgyay, twin brothers and architects, wrote several articles and books on "bioclimatic" architecture—building design that takes environmental conditions into account. Their articles on "sun orientation" and "environment and building shape" explained a new type of architecture, designed to capture energy from the sun, and one with scientific underpinnings.

Importantly, Victor Olgyay created a bioclimatic chart that shows how human comfort is really dependent not just on temperature but also on other weather factors, including the effect of air movement, humidity, and radiation. Comfort, he proved, was not solely a matter of cooling temperatures. It could be achieved through natural means as well. For instance, people can also feel comfortable through a breeze, even when the room temperature may be relatively warm. This offered an alternative to air-conditioning, such as natural ventilation.

The most futuristic of solar home creations came from the hands of American architect and inventor Buckminster Fuller. Out of frustration with inefficient homes, in the 1940s he designed a mobile mass-produced house, the Dymaxion Dwelling Machine, with an environmentally friendly footprint. Shaped like an aluminum spaceship, it was heated and cooled by natural methods. A vent protrudes from the domed roof, shaped to induce a heat-driven vortex sucking air. It provided the entire house with natural ventilation. Despite its space-age aesthetics, few people bought a Dymaxion dwelling.

Fuller, best known for his geodesic domes, also investigated ways to make entire cities more energy efficient through massive domes. In 1960, he proposed a gigantic dome over Midtown Manhattan, about two miles in diameter. Helping to correct the wasteful nature of cities, it would work like a large greenhouse, keeping heat inside the dome during the winter, while shielding residents from snow and rain. The dome would make for a high-efficiency city under this constant climate, saving money on heating, and also on umbrellas. "The cost of snow removal in New York City would pay for the dome in 10 years," Fuller noted.[26]

Fuller's grandiose vision, bordering on megalomania, was never executed. However, his idea of a dome-covered city, as bizarre as it may have seemed, made a lot of sense for an extremely cold climate. Eventually, a reduced version with a 160-feet diameter was built in Antarctica: the Amundsen–Scott South Pole Station.

The Italian-American architect Paolo Soleri developed equally large visionary structures called "arcologies," a portmanteau of architecture and ecology. For decades he drew up densely populated habitats with a low environmental footprint. His 1960 plan for Mesa City was meant to house 2 million people in a series of organic-looking megastructures, like 750-foot-high termite mounds.

Despite Soleri's many scroll drawings of the project, some as long as 180 feet, it was never realized. Nor was his vision for Hexahedron in 1969, a more than 1 km tall human-made mountain, composed of a pyramid and an inverted pyramid, encrusted with landscape. This beehive of human activity would contain an entire city of 170,000 people within a single structure. Homes were to be placed inside the top pyramid. Carved into the bottom pyramid would be a cultural center. The giant legs holding the pyramids would contain offices.

Soleri ended up building his own toned-down version in Arizona, called Arcosanti, founded in 1970. Its better buildings are shaped like domes and arches to provide shading. The buildings rely on passive solar strategies, including buildings with thick walls to benefit from "thermal mass": the capacity of materials to store heat. Constructed with minimal resources, it was supposed to accommodate 5,000 people. It rarely exceeded 150. By 2008, growth stalled. "I don't have the gift of proselytising," Soleri stated—although his more serious faults were exposed upon his death, including egocentrism and sexual abuse.[27]

While few of Soleri's and Fuller's grandiose visions became a reality, they inspired a new generation of architects to reduce architecture's imprint on the environment, including those designing skyscrapers. British architect Norman Foster, who collaborated with Fuller for over a decade, was able to turn environmental idealism into tangible and successful buildings. In 1985, he created the HSBC Building in Hong Kong, pioneering a more sustainable tall office building. Structurally, the building appeared like an offshore rig, with bracing on the outside. With the structural needs of the building solved by the exoskeleton, he could innovate in the building's interior. Whereas the typical tower has an elevator core on the interior, Foster moved the core to the side, allowing for a spacious interior atrium. He installed mechanical "sunscoops" on the south side of the building. These banks of giant mirrors reflect

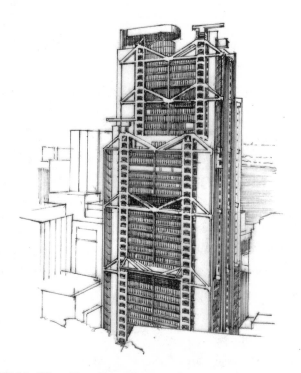

HSBC Building, Hong Kong, Foster and Partners, 1985

natural sunlight into the building, moving with the position of the sun throughout the day. The reflections aim to reduce the necessity for artificial lighting and the extra cooling required by the electric lighting. While his building still required air-conditioning, it pumped seawater, instead of fresh water, as coolant.

In 1997, Foster completed the Commerzbank headquarters in Frankfurt, Germany, then the tallest building in Europe, at 985 feet. Strict German environmental law, coupled with Foster's drive for more sustainable architecture, created an unprecedented project. Like the HSBC Building, Foster created a central soaring atrium, which naturally acts as a chimney to remove warm air. He also created spirally dispersed indoor winter gardens throughout the building, allowing light to penetrate deep inside the tower. Departing from the hermetically sealed box, he specified facades with operable windows enabling natural ventilation. Foster showed that a more sustainable skyscraper could also be a pleasant place to work, filled with vegetation and spacious gardens.

As much as Foster's skyscrapers presented a new future for tall buildings, his revolutionary layout with lots of open atria scared leasing agents looking to maximize their floor-space rents. Only a few visionary clients would be willing to take this risk. How could the idea of more sustainable tall buildings become widespread?

Around the same time as the Commerzbank opening, in 1996 in Darmstadt, a half hour from Frankfurt, the Passivhaus Institute was founded. A German physicist, Wolfgang Feist, created a new voluntary building criteria standard for ultra-low-energy buildings, called the Passivhaus standard. Dr. Feist had taken an interest in a handful of buildings that were built in the 1970s, in response to the oil embargo, such as the Saskatchewan Conservation House, built in 1977. Canadian researchers created this demonstration house to prove better ways to build homes. They found that a typical building could barely keep any heat inside. The average home of that period lost 30 percent of its energy through air leakage, 25 percent through the basement, 15 percent through the ceiling, 15 percent through the walls, plus 15 percent through the doors and windows.[28]

In response, the Canadian researchers would build their homes with extremely thick insulation—more than six times the required amount—and without a furnace. There would be no "thermal bridges," preventing energy from leaking through. Their homes also featured triple-glazed windows and were virtually airtight, making sure not an ounce of energy was wasted. They even included a novelty: a heat-recovery ventilator, which reuses the heat of outgoing, stale air to warm up the cooler, fresh air. Unlike Foster's towers, their homes were modest, a few stories at most, and nearly forgotten. But today they, along with the researchers of a few other energy conservation homes such as the 1976 "Lo-Cal" (low calorie) house at the University of Illinois, are held in high esteem for their work as "super-insulation" pioneers.

The Passivhaus Institute has helped these energy-saving innovations become more mainstream. In addition to requirements such as energy-recovery ventilators, the institute passes its certification on structures only when they pass the "pressure test." Instead of air-conditioning, Passivhaus designers want airtightness. This led to manufactured blower doors in the 1980s, large fans mounted in doors and window openings to measure the airtightness of rooms. The fans blow air into or out of the room, creating a pressure differential between inside and outside. The more airtight the building envelope and its air ducts, the less air is needed to create a change in building pressure.

By 2010, about 25,000 Passivhaus-certified structures had been built. They were mostly a few stories each. The Passivhaus standard would not work for skyscrapers, critics thought. They pointed to tenants paying extra for large floor-to-ceiling windows, even though windows are poor insulators compared to walls. In addition, they gasped about the inefficiency of floor plates as a result of the very thick insulation of Passivhaus requirements. For these reasons, they thought, the Passivhaus would stay precisely that—a *Haus*, not a *Wolkenkratzer*.

To their surprise, in 2013 a 260-foot-tall office building, called the RHW.2, was completed in Vienna that followed Passivhaus standards. It featured a double glass wall with an inner cavity, allowing people to open the inner windows for fresh air without major energy loss. The tower managed to keep a glass aesthetic while the extra skin dropped the energy usage to only 20 percent of a typical building. While this double facade cost a few million dollars extra, the lower energy bills would recover the cost in only four years.[29] The project kicked off a race for the world's tallest Passivhaus skyscraper.

While the tower drastically reduced the building's operational energy, it did not address glass's embodied energy, the energy created to make the material. Glass is a very energy-intensive material with a relatively short life span, requiring replacement every thirty to forty years. And recycling is complicated, with glass panels attached to each other with plastic.

The problem lies in our obsession with the crystalline aesthetic. Glass may be great for windows, but for other parts of the building, it cannot compete with more sustainable materials.

In 2018, Cornell Tech created The House in New York City, a residential skyscraper for faculty and students to set the record with a 270-foot-tall Passivhaus. Instead of all glass, the high-performance facade consists of both glass and prefabricated metal panels.

Departing from the all-glass aesthetic, it became a poster child for how to turn New York into a Big Green Apple.

More New York towers will follow this example, thanks to the 2019-enacted Local Law 97, the world's most ambitious climate legislation for large buildings. It puts buildings larger than 25,000 square feet (about 50,000 buildings in New York) on a path to meet the city's goal to reduce carbon emissions by 80 percent by 2050. It does so by forcing owners to cap the carbon use per building. If they don't comply, they will end up paying substantial fines that could run into the millions of dollars.

While a marvel of environmental performance, the Cornell tower, with its narrow bands of windows and gray panels, aesthetically lacks the luster of a typical glass tower. Creative designers can make a difference here. The same year, across the Atlantic Ocean in Bilbao, Spain, a Passivhaus tower called the Bolueta was completed. It was a little taller, at 289 feet, with a facade having a lot more pizzazz. Instead of relying on glass for its prism aesthetic, architect Germán Velázquez angled glossy surfaces to enliven the building's exterior. The color is black as coal, a reference to the city's industrial past, the coal-based industry. As light reflects from the surface, the building sparkles, and feels like a nod to the future.

Today, taller and even more aesthetically pleasing towers are in the works. In Vancouver, a Passivhaus skyscraper is planned for 586 feet. Since 2016, the city itself is rezoning buildings to meet operational energy limits almost similar to Passivhaus standards.

The rules of the game are changing. In an age of climate change, height alone is no longer a proper measuring stick. Our tallest buildings can no longer be monuments to Carrier's invention and Bauhaus aesthetics.

Skyscrapers need to become lean, airtight, Passivhaus machines.

PASSIVHAUS MAY BE only a voluntary standard, but with cities tightening their energy building codes, it may be coming to your house as well. Chicago, the birthplace of the skyscraper, launched an energy benchmarking system for large buildings, forcing them to report their energy use to the city. The buildings must then display public placards with their energy performance, from a meager one-star to a four-star rating. The hope is to increase awareness and encourage efficiency of a building's green initiatives while putting the lower-performing skyscraper owners on notice.

Even architects' design work may end up being rated. Professional institutes around the world, such as the American Institute of Architects, have joined the 2030 Challenge, which commits architecture firms to design carbon-neutral buildings. Architecture firms that want to join will have to submit their annual portfolio of buildings in the design phase, including statistics to predict baseline energy performance. This will give them green bragging rights. Soon, there may be a Michelin Guide for hiring a sustainable architect—or the opposite, a blacklist of environmentally wasteful buildings, to deter you from renting an unsustainable office space.

With more efforts aimed at reducing our reliance on air-conditioning, will this mean the end of glass buildings? Is the glass skyscraper a monument of the past like the Bauhaus, an inefficient monument of an age of coal abundance?

On top of being an energy waster, glass is also a serial bird killer. Sadly, a shocking thump you hear on your glass window is a lot more common than people think. Up to one billion birds are killed each year in the United States alone through such collisions.[30]

Years of lobbying by bird advocates, rehabilitators, and experts have led to the 2019 passing of a New York City mandate of "bird-friendly" glass in new buildings. It requires that the first 75 feet of new skyscrapers use material visible to birds. Windowpanes should allow birds to see inside the buildings, instead of glass mirroring the clouds. Ceramic frit glass, created by screen printing glass with enamel in a small dot pattern, helps our feathered friends more easily recognize the windowpane as a curtain wall. Glass could also integrate ultraviolet-reflective coatings visible only to birds but not to humans.

Building glass as we know it is changing in profound ways. "Smart glass" now comes with its own IP address and an ability to automatically adjust to glare and sunlight, or be controlled with a smartphone app. When a window is shattered, it can even call the police. Electrochromic glass panes can change tint several shades from opaque to transparent by running an electric current. The darker tints can block radiation and reduce energy costs related to cooling. In the future, glass may even be a screen for video to project on—a yet untapped place to feature advertising.

Smart glass is more expensive. As it makes otherwise unusable, overlit space fit for workers, it may help accommodate more people in the building and end up being more efficient. Unlike smart doors, there is less of a technological risk. It's unlikely that a smart window is something a cyber criminal may want to hack, unless they figure the best way to wreak havoc is by controlling your window tint.

Glass panes can also incorporate photovoltaics and generate electricity. Already, glass houses exist with pink-tinted glass. This lets in only sunlight's wavelengths that actually stimulate plant growth,

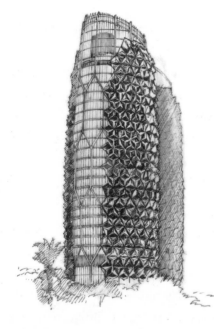

Al Bahr Towers, Abu Dhabi,
AHR, 2012

while others are converted to electricity. Photovoltaic windows with translucent photocells exist as well. These are one of a host of building-integrated photovoltaic products, including roofs and skylights.

While our glass is becoming smarter, so are our external window blinds. Thanks to advances in robotics, external shading devices can become responsive, leading to full-fledged kinetic facades. The 475-foot-tall Abu Dhabi Al Bahr Towers feature a shading facade containing many star shapes, inspired by a traditional Islamic lattice shading device. These kinetic stars track the sun's path. As the sun moves, the petals of the stars open and retract to ensure appropriate shading.

These kinetic facades may make sense for high-sun environments, although probably not yet for your average building, given their high cost. Instead of mechanical solar shading, which is prone to breaking, there could be a solution that relies on the intrinsic properties of the material to trigger a shape change. For instance, wood expands as it absorbs water. Researchers are already looking to create a material that moves or expands as the sun moves or the temperature increases. Shape Memory Alloys, for instance, change shape when cold and return to their pre-deformed shape when heated. Through biomimicry, these researchers take their inspiration from the biological processes in nature, such as plants adjusting their leaves to track the position of the sun.

While our future responsive building envelopes may require less cooling inside the building, it's unlikely that air-conditioning systems will be going extinct anytime soon. Your old-fashioned air conditioner is undergoing fundamental changes, including getting a virtual avatar. Machine learning and "smart" devices have led to a technology called a "digital twin." This is a virtual replica of a building, the bridge between the physical and the digital world. Intelligent control mechanisms can mine real-time data from sensors in the building, such as air quality and people's activities. Artificial intelligence algorithms run thousands of simulations to test new air-conditioning settings or to uncover problems the real world can benefit from.

Still, the best way would be to require virtually no air-conditioning at all. Again, architects have a thing or two to learn from nature. A mid-rise office and shopping complex in Zimbabwe, Eastgate Centre, which opened in 1996, managed to cut its energy use to only 10 percent compared to a similar-sized air-conditioned building, according to the architect, thanks to a natural cooling system inspired by local termites. Long before humans existed, termites were students of passive cooling.[31] Inside one type of termite nest, termites farm a fungus that must be stored at a constant 87 degrees, even as the desert temperatures swing dramatically from daytime heat to nighttime cool. They achieve this through a system of vents and flues. During the day, they vent breezes from the cooler base of the earthen mound up through a flue at the top. The termites tirelessly open and close the vents to enhance ventilation in order to keep their only food from spoiling.

These efforts may be promising ways to help ourselves get out of the situation we ourselves created. Business as usual, with ever taller curtain-wall buildings, will only increase our reliance on air-conditioning and further contribute to global warming.

To avoid living in an air-conditioned world forever, it's time we cut the cord, and rethink our glass towers.

PART II

SOCIETY

FIVE

The Rules That Shape Skylines: London

For three centuries, London's skyline was defined by St Paul's Cathedral, with its famed silhouette of Sir Christopher Wren's dome and spires. Today, the horizon is crowded with skyscrapers of increasingly eclectic shapes that have earned them their nicknames: the Shard, the Gherkin, and the Cheesegrater.

Common opinion once held that skyscrapers had no real place in European cities, with their historical buildings and mature economies. Skyscrapers were reserved for cities most heavily destroyed by bombing in World War II, like Rotterdam or Frankfurt. Until recently there has been no love lost between Europe and tall structures. Even Paris was going to do away with the Eiffel Tower. When Gustav Eiffel proposed his namesake tower in the nineteenth century, the city's cultural elite dismissed it as a "gigantic black smokestack." Upon construction, the city seriously considered demolishing it and selling it as scrap metal. To avoid this fate and to prove the hollow tower's utility, Eiffel himself erected an antenna and financed wireless telegraphy experiments. When the French military recognized the tower's newfound usefulness, a Parisian committee with authority for the structure's fate only reluctantly let it stand ... and ambivalently. "One would wish it were more beautiful,"[1] the committee stated.

From the nineteenth century to the present day, only one new skyscraper was built in the center of Paris. The 689-foot-tall Tour Montparnasse, completed in 1973, became known for being "the most hated building in Paris."[2] It stood out so much over Paris's historic buildings that all future skyscrapers were banished to La Défense, several miles outside of the city center.

Technology might make skyscrapers possible, but all of this may be futile if society does not want them.

Yet today, central Paris counts three new skyscrapers, each of them out-"Eiffeling" the Eiffel in their own way. They are the Tour Triangle, dubbed the Toblerone for its triangular wedge shape; the Tribunal de Paris, a series of stacked blocks; and Tours Duo, a pair of drunken leaning towers. At the same time, a veritable skyscraper boom has been happening across the border in Switzerland, known for its mountain peaks and pine chalets—not high-rises.

In the postwar era, urban populations in Europe and the United States declined, as people moved to the suburbs. Since the 1980s, there has been a reversal of the suburban exodus. Never before have more people chosen to live in cities. In the United States, Europe, and Asia, younger and more highly educated people are moving to cities, attracted by the opportunities the cities create. Urban planning researcher Markus Moos has called it the "youthification" of American cities, with two-thirds of twenty-five- to thirty-four-year-olds with a bachelor's degree living in metropolitan areas.[3] Similarly in Europe, large cities like London have been magnets for people in their twenties.[4]

This millennial generation, unlike their parents, is far less drawn to home or car ownership or to a car-oriented lifestyle. One survey found that 30 percent of American millennials do not plan to buy a car at all.[5] Millennials are more inclined to spend money on travel, dining, and concerts. They would rather live in cities, where such potential life experiences are within easy reach.

Now all of this back-to-the-city migration is constraining city land supply and leading to higher land prices. Since skyscrapers maximize the number of people on a given plot of land, they can also offset higher land costs. When investors realized these tall possibilities, they had to find a way.

Before developers can weave a new colossus into the finely woven historic fabric of old European cities, they must overcome significant regulatory hurdles. Building regulations and zoning, enacted to protect the public and property owners, have long hindered tall buildings.

For decades, London's city officials blocked skyscrapers. Londoners did not even want to name them that—preferring the term "tall buildings." Today, demographic winds and tastes have changed. London city planners now allow tall buildings, if they and the public are swayed by the lure and elegance of the proposed building shapes. In Paris, city officials eager to bring potential for job growth approve towers, even in the face of vocal opposition.

Historic skylines are now permanently changed. London's skyscraper boom created an entirely new skyscraper cluster, as well as Western Europe's tallest tower, the Shard. This slender, glass-clad pyramid is the most potent example of the advent of supertalls in the Old World. If you squint hard enough, its tapering crystalline aesthetic almost seems like a slightly more modest version of Wright's Mile High tower.

As the new towers keep coming, changing centuries-old skylines

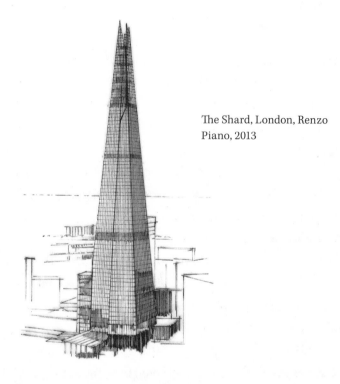

The Shard, London, Renzo
Piano, 2013

for the sake of expensive offices and luxury penthouses, many Europeans wonder if they have gone too far. All these high-rises summon existential questions for European cities. Should they protect their heritage at all costs and keep their cities frozen in time? Or should they allow new buildings, with the consequence of altering their picture-perfect skylines?

CITIES HAVE ALWAYS BEEN magnets for people and engines of civilization. Anthropologists consider urban centers a core characteristic of civilization, on par with agricultural methods, written language, government, standards of measurements, and a shared culture. In fact, the words "city" and "civilization" have the same

Latin root of *civitas*. This once meant a community but eventually came to mean a city in a physical sense.

It wasn't until our ancestors developed agriculture that they were able to leave their nomadic lifestyles and live in permanent settlements. With agricultural surpluses, they had the frame of mind to focus on more than just hunting and gathering. Cities became the breeding grounds of new professions and specializations. With their primal physiological needs satisfied, urban societies progressed up the metaphoric pyramid of human needs—and as a result, they built pyramids, literally.

The city may have led to civilization, but it also led to the rule of law in order to counteract its potential for trouble. With people living close to each other, disputes arose. The ancient Greek word for city is *polis*, which has the same root as "politics." "To be political," wrote political theorist and philosopher Hannah Arendt, "to live in a polis meant that everything was decided through words and persuasion and not through force and violence."[6]

Words and persuasion led to laws. Soon, the rule of law turned not just to crime but also to building violations. The first known written building regulation dates from about 1754 BC, the Code of Hammurabi, after the sixth ruler of Babylon's First Dynasty. The code spanned the full spectrum of everyday disputes, from the wages to be paid to an ox driver to the punishment to be meted out to anyone who stole a sheep or a goat. The law worked with exact reciprocity, following the "eye for an eye, tooth for a tooth" principle.

"If a builder builds a house for a man and does not make its construction firm and the house collapses and causes the death of the owner of the house—that builder shall be put to death. If it destroys property, he shall restore whatever it destroyed, and because he did not make the house firm he shall rebuild the house which collapsed at his own expense."

In addition to building regulations, ideas around city layouts also took hold. Leaving people to build their homes anywhere they wanted could lead to an uncoordinated whole. Hammurabi had already attempted to formally organize the city through ordering the rebuilding of the maze of Babylon with wide, straight streets and paved with bricks.

Efforts at city planning date back even further. Ancient Egypt, with few records of organized settlements, was once seen as the exception to the norm of civilization following cities. New evidence points to well-planned cities in the Fourth Dynasty of the Old Kingdom around 2,500 BC.[7] Remains of these settlements reveal that they were part of a larger plan. Homes were neatly organized around rows in identical layouts and along straight streets and at right angles. They were planned in grids. Egyptian pharaohs, known for their pyramids, may have also been the world's first urban planners.

Soon, ideas around urban planning would determine the shapes of many cities. The ancient Greeks popularized more formal city planning, although grid plans existed in other societies as well—such as the twenty-sixth-century BC Indus Valley and fifteenth-century BC China. In the fifth century BC, Hippodamus of Milete, known as the "father of urban planning," designed the grid system of Piraeus, near Athens. His "checkerboard" plan featured straight streets crossing at right angles. He believed the grid expressed the rationality of civilized life.

Aristotle credited Hippodamus not only as the inventor of "the division of cities into blocks" but also with the first land use plan. He divided the city depending on class into three parts. "One sacred, one public, and one private," Aristotle wrote. "Sacred land to supply the customary offerings to the gods, common land to provide the warrior class with food, and private land to be owned by the farmers."[8] This first land use plan was codified by Plato's and Aristotle's laws in the fourth century BC, and then was further developed. They required each town to have a public square, or *agora*, the thriving hub of Greek society and the birthplace of democracy.

The Romans borrowed many city planning principles from the Greeks, and added new levels of detail that spread across their empire. Vitruvius suggested that the public square, or Roman *forum*, should be rectangular, with a 3:2 length to width ratio, and its size proportionate to the population. It should be in the center of the town if inland, or near a harbor if in a coastal city. This law later influenced the Laws of the Indies, and largely determined the fabric of cities in the Americas.

Vitruvius wrote about other city planning principles as well, including where to build a settlement. His writings showed Romans obsessed over picking a "healthy" site. "Such a site will be high," Vitruvius wrote, "neither misty nor frosty, and in a climate neither hot nor cold, but temperate; further, without marshes in the neighborhood. For when the morning breezes blow toward the town at sunrise, if they bring with them mists from marshes and, mingled with the mist, the poisonous breath of the creatures of the marshes to be wafted into the bodies of the inhabitants, they will make the site unhealthy."[9] He even strongly insisted on a return to "the method of old times," when our ancestors, before building defensive works, examined the livers of sacrificed cattle to determine the quality of the local water and food supply.

Vitruvius's writings cover where to place important buildings, plus how to lay out a city to optimize for defense or climate. He also prescribed building standards, including types of materials, wall thickness, height, and the relationship between buildings.

The Romans regulated the uses of cities as well. In Imperial Rome, rules restricted operating industries that could cause harm, such as cheese smokeries and tanneries. Brickyards and cemeteries were confined to the outskirts of the urban populated areas. As the empire spread, Roman military camps were laid out with divisions, locating *principia* (main quarters), *fabrica* (workshops), and *horreia* (granaries), among other functions. These military rules, while initially only for temporary settlements, would have long afterlives, when military camps evolved into cities.

The Middle Ages brought medieval communes with charters establishing municipalities and regulating aspects of urban life. A count or a duke would grant a settlement town privileges, allowing a package of rights, which in some cases included the right to call itself a "city." The charters gave various rights such as to construct a city wall, operate a central market, and levy tolls. They even gave cities the right to create their own laws. This would lead to increasingly detailed building regulations. Some of these controlled the relationship between private property and public assets, such as the size of building colonnades or the maximum protrusion of buildings near city walls. These may have otherwise compromised the city's defense.

An early precedent of tower regulations dates from this age. In the fourteenth century, two rival families in the Italian town of San Gimignano competed for the tallest tower-house, leading to masonry structures taller than 200 feet. The council intervened. They ordered that no structure should exceed the tower attached to its seat of power, the *Palazzo Comunale*. (Legend has it that, in response, one family erected *two* adjacent tower-houses, each only slightly shorter than the maximum height.)

With urban growth and densification, risks grew as well. In the eleventh and twelfth centuries, several fires devastated London.

City officials wanted to encourage people to build houses in stone. In 1189, London's first lord mayor set up the Assize of Buildings. This expanded the scope of regulations. Owners were recommended to build fire-retarding barriers between their houses, "at their joint cost, a stone wall three feet thick and sixteen feet in height."

Unsanitary conditions and overcrowding became common in European cities during the Middle Ages. These partially resulted in high infant mortalities and pandemics such as the Justinianic Plague. In response, city officials started to regulate sanitation and stormwater flows. In addition, the elite dedicated themselves to urban sanitary improvements. In one district in thirteenth-century London, an English queen was remembered for her role in building the first public latrine.[10]

The roots of a major obstacle for future tall buildings dates to this age. The assize started to regulate views: "If any person shall have windows looking upon his neighbour's land, although he may have been for a long time in possession of the view from such windows, and even though his predecessors may have been in possession of the windows aforesaid, nevertheless, his neighbour may lawfully obstruct the view from such."

As long as new development did not encroach on a neighbor's land, builders would not have to consider views. Without the right to overlook someone else's property, builders could do as they pleased—at least until the tide swung the other way.

When Europe's major cities became overcrowded in the fifteenth and sixteenth centuries, health problems propelled further regulation. Construction in London occurred beyond the city's walls, where sanitary facilities were lacking. Elizabeth I passed a decree making it illegal to build within three miles of any of London's gates.

On September 2, 1666, a fire from Thomas Farriner's bakery on

Pudding Lane spread easily in the narrow streets from warehouse to warehouse, many of them containing flammable substances such as timber and oil. The Great Fire of London lasted five days, destroyed one third of London, and made about 100,000 people homeless. It led to the Rebuilding Act of 1667 and further regulation of buildings. The government prohibited narrow alleys and mandated standards for streets. Buildings were classified into strict categories depending on the street they faced. "There shall be only four sorts of Buildings and no more," the act stated. Each type was to follow rules about wall thickness, building materials, and overall height. "The Number of Stories, and the Height thereof, be left to the Discretion of the Builder so as he exceeds not four Stories." The city even appointed surveyors who were tasked with enforcing the codes and were empowered to jail violators. Thus was the building inspector born.

During the Renaissance and the Baroque eras, building regulations became more comprehensive. They came to include the cultural elite's pursuits of universal beauty and rationality. They mandated aesthetic elements, in addition to safety. In seventeenth-century Berlin, officials prescribed everything from the house number to the building's color. In the suburb of Friedrichstadt, it was a very light gray, uniformly applied.[11] Officials issued work permits only after they deemed the builders' plans and colors appropriate and tasteful—making them the precursors to the modern day co-op board and home ownership association.

With the Industrial Revolution, even broader regulations took hold. Growing urban populations along with the construction of factories inside of cities had increased pollution and led to dilapidated conditions. From 1800 to 1860, England's urban population grew from 17 to 72 percent.[12] Friedrich Engels described Manchester as evoking "Hell upon Earth":

"I am forced to admit that instead of being exaggerated, it is far from black enough to convey a true impression of the filth, ruin, and uninhabitableness, the defiance of all considerations of cleanliness, ventilation, and health. . . . If any one wishes to see in how little space a human being can move, how little air—and *such* air!—he can breathe, how little of civilisation he may share and yet live, it is only necessary to travel hither."[13]

By 1900, most of London's then five million people lived in these horrendous conditions. A lack of sanitation led to cholera and typhoid, plus other diseases. Smog and pollution enveloped the city in a permanent shroud, leading a Victorian poet to coin the nickname "the city of dreadful night."[14]

The industrialization of cities got in the way of civilization. In the late eighteenth century, the urban bourgeoisie began to flee to villages and to create landscape gardens outside the city of London—the first form of modern suburbia. Ironically, these areas initially attracted prosperous businesspeople who had made their wealth from the Industrial Revolution but then had fled the smog and pollution it created.

In contrast to their British and American counterparts, the Parisian bourgeoisie did not flee to the suburbs. They decided to reshape their city. Napoleon III enlisted Baron Haussmann to clear the slums and relocate them to the fringes of the city. Haussmann's 1850s plan included wide boulevards cutting through the medieval street plan of old Paris. Strict rules regulated building height, roof lines, materials, monuments, and building facades. In 1883, a new law even prescribed dimensions for columns, pilasters, capitals, cornices, and virtually every decorative element.[15]

Haussmann's considerations went beyond aesthetic and sanitary rationale alone. The narrow medieval streets, the terrain on which the French Revolution had been fought, had been easy for rev-olutionaries to barricade. So, the straight lines of Haussmann's wide boulevards would make the city easier for the military to navigate and for artillery to fire at the crowds. The long boulevards leading to major public monuments would help visually to instill citizen-ship in the minds of Paris's inhabitants, with a dozen boulevards converging at the Arc de Triomphe, leaving pedestrians with a clear reminder of the empire's might.

London's 1844 Metropolitan Buildings Act exceeded even Napoleon III's level of detail. It established the Metropolitan Build-ings Office to regulate buildings, and categorized them into three classes, including dwellings, warehouses, and public buildings such as schools, churches, and theaters. Each class of building had its own set of elaborate rules, and government surveyors were empow-ered to enforce them. Certain activities were expelled from the city altogether, such as those likely to cause fire or generate noxious effluents, like boiling blood.

These specialized buildings stood in contrast to ancient build-ings, which were mostly the same for each use. Few structures were specialized to accommodate an industry, for instance. "The House," noted one researcher about ancient Greek buildings, "was not iso-lated from work."[16]

During the Second Industrial Revolution, taller structures disrupted urban life, which spawned more regulations. In late nineteenth-century London, then the world's largest city, a block of Westminster flats rose taller than 100 feet. Several people com-plained, including Queen Victoria. The tall building blocked her view over Parliament from Buckingham Palace. It triggered the 1894 London Building Act, which suppressed the height of buildings, ensuring that the city's landmarks, in particular St Paul's Cathedral, would not be obscured. However, the prime motivation of the acts was actually fire safety, with tall buildings not to exceed the height of a fire escape ladder at 80 feet.

So, the tug of war between high-rise developers and regulators began. In 1933, the Unilever House managed to exceed the city's height limits. Its builders had found a loophole in the regulations, claiming their top floors did not need to satisfy the height limits since they were for neither office nor residential use.[17]

This led to a public outcry over the newly obstructed views of St Paul's dome. The Royal Fine Arts Commissioners commented on how the building, as well as another offender, had "disastrously blocked some of the most famous and beautiful prospects in London."[18]

In response, architect W. Godfrey Allen surveyed all the remaining views of St Paul's. He outlined eight "protected view corridors," based around multiple views of the cathedral and Westminster from special vantage points in the city, such as from major parks like the General Wolfe statue in Greenwich Park and the summit of Parliament Hill. In 1938, the City of London Corporation inaugurated the city's "protected views" system. No longer would tall buildings be permitted to ruin the view of St Paul's cherished dome and spires.

During the reconstruction of the city after World War II, this regulatory framework protected London's views, albeit from Allen's chosen vantage points. City officials added more views, including to protect the vista of the Tower of London plus the River Thames. Together, the thirteen protected views ensured that London remained a low-rise city. Skyscraper developers would have to fight a list of authorities including the mayor of London, four London planning bodies, and English Heritage, the custodian of historic places. In addition, developers had another hurdle to overcome: a regulation known as the "right to light." The "Doctrine of Ancient Lights" originated from a 1663 property law that would come to protect "Ancient Lights." Buildings that had received natural light through their windows for twenty years or more retained the right to this view forever, impeding new developments.

Up until 1832, homeowners declared this right by painting the words "Ancient Lights" underneath their windows. Some of these can still be seen on London's oldest residential buildings. Today, rights to light are automatic. Owners can exercise this right by proving that a proposed building would breach their property's right to light. This requires a careful calculation, based on a somewhat arbitrary standard. In the 1920s, Percy Waldram and his father assessed the minimum sufficiency of light, the so-called grumble point. They defined this in *The Illuminating Engineer* as "the natural illumination at which average reasonable persons would consistently grumble."[19] We now consider having natural light as a bonus in our homes, but through most of history it was essential. Based on their research, they found this "grumble" threshold to be at 0.2 percent of the entire sky level—a minimum of one lumen per square foot. This is also known as one foot-candle, the amount of light from one small flame shining on a one square foot surface from a foot away, on a dreary day in winter.

If a neighbor infringes on this right to light, even by planting a single tree, owners have the right to sue for "nuisance" in order to get a sizable check, with hopefully a restored view. Developers who ignore these rights do so at their own peril. At the cost of a lot more than just a grumble, they are potentially forced to demolish their building by removing the offending part.[20]

Together, the protected views and rights to light would impede high-rises during the twentieth century. St Paul's Cathedral remained the tallest building for more than 250 years—except for the 1962 Post Office Tower, which is technically a TV tower, not a skyscraper. Still, when height restrictions were lifted in 1956, few true skyscrapers emerged in central London. Most of them were built at Canary Wharf, the redeveloped Docklands, far east of the central city.

Across the Channel in Paris, there was equal resistance to sky-scrapers, if not more. There had been no skyscrapers since 1973, with the completion of the 59-story Tour Montparnasse, for the longest period the tallest tower in France. It did not inspire oth-ers, especially since it was voted by one international poll as the world's second-ugliest building.[21] Most buildings were between five and eight stories to fit the character of boulevards, following Hauss-mann's nineteenth-century plan.

In London, developers mounted pressure on the view corri-dors, some of which originated from more than ten miles away. For instance, King Henry's Mound, one of the protected view vantage points, is so far from St Paul's that it is hard to actually see the dome, even if you use a telescope. The roots of this protected monument, in the name of heritage preservation, were questionable in their own way. Christopher Wren, the renowned English anatomist, astrono-mer, physicist, and architect, had ignored existing heritage to rebuild St Paul's after the Great Fire of London of 1666. Ironically, he had used gunpowder and a battering ram to clear the site of its histori-cal remains. If he had had his way, his plans for a post-Fire London would have reconfigured the city with long, wide boulevards dotted with classical buildings, not unlike Haussmann's Paris.

In the new millennium, with urban populations once again growing, the pendulum would swing back to developers. As soon as city officials in Europe's oldest cities opened the regulatory road, time ran out on flat landscapes. Their skylines filled with skyscrapers.

IN 2000, the Italian architect Renzo Piano, tasked with designing London's largest building, drew inspiration from the city's nautical history. He created a spire shape rising from the River Thames, comparing it to "a 16th century pinnacle or the mast of a very tall ship."[22]

But the English Heritage group had a different interpretation of the architect's proposed project, a slender glass pyramid. They criticized the tower for resembling a "shard of glass through the heart of historic London." The name stuck as the "Shard."

The group feared that the Shard would radically alter the experience of central London. Previously, a walk through London's medieval streets or along the banks of the Thames would give you sights of St Paul's classical dome or the ornate Victorian Gothic details of Tower Bridge. With the Shard, "New London" would loom on the skyline, rivaling the old with its glassy, glossy, and contemporary shape.

At least Piano wasn't going to design the Shard as a run-of-the-mill skyscraper. Billed as a "vertical village," it would include 26 floors of office space, three floors of restaurants, 19 stories of luxury hotel, 13 stories of apartments, and an observation deck. Mixing these uses made sense with the structure's tapering shape. The wider bottom floors are a good fit for the larger office spaces, with the narrow top floors more appropriate for residences.

However, this required the Italian architect to fulfill the different requirements of each of these uses together in one building, when each use has its own needs. For instance, the couple taking the elevator to their honeymoon suite at the Shangri-La may not want to share their elevator cabin with an army of office workers.

Arguably, Renzo Piano's biggest challenge was to get his pinnacle through regulatory hoops. Fortunately for him, the groundwork had been laid by another skyscraper. Across the Thames, at 30 St Mary Axe in the City of London, rose another tower. Upon completion in 2003, it took the place of the Baltic Exchange after the IRA had detonated a one-ton bomb outside the historic building. Architect Norman Foster had proposed a 590-foot-tall tower in the shape of a stretched egg with a distinctive diagrid lattice, leading to its nickname, the Gherkin. Even though the building wasn't in any of the protected view corridors, it required the deputy prime minister's approval to allow for a building much taller than the old Exchange.

30 St Mary Axe, London, Foster and Partners, 2003

Where city officials for decades had blocked tall buildings, a new administration had more favorable opinions on high-rise construction. Ken Livingstone, the mayor of London, kick-started a new London plan. "For London to remain a competitive world city," he wrote, the city "must respond to the drivers of growth ... without inappropriate restraint."[23] Livingstone would eventually green-light fifteen skyscrapers.

The Gherkin, lauded by architects and the public alike, changed Londoners' minds about tall buildings. It opened the floodgates for even taller buildings, like the Shard. Unlike the Gherkin, the Shard was planned to be inside the view corridor of St Paul's Cathedral from Kenwood House. English Heritage appealed the Shard's plans, questioning the design of the tower and its "oppressive" impact on the skyline. The Shard was subjected to a strenuous public inquiry process.

Piano needed to limit his tower's potentially overwhelming impact on the city, plus ensure that its five thousand workers would not clog up any roads. Fortunately, his site stood right next to the major transit interchange of London Bridge Station. Commuters could easily take public transit, avoiding the problem of more parking and cars.

Piano also had to manage the difficult contradiction of making the massive structure seem smaller. His tower would "melt with the City" at the bottom, its glass opening up at the tower entrance. At the top, it would "come to almost nothing,"[24] with the pyramid nearly converging to a single point. He angled the glass of his pyramid so it would reflect the sky and sunlight instead of neighboring buildings. As a result, it would change its appearance with the changing weather and seasons, plus be less obtrusive on the skyline. His tower would not block the view of St Paul's from Hampstead Heath, since it's located behind the dome. The 1,016-foot-tall glass colossus—almost three times as large as the cathedral—would, however, change the view quite a bit.

Architect Richard Rogers, who served as a chief consultant to the Greater London Authority, defended the tower. The tower would allow for a new dialogue that would make the dome better off, he argued. "The contrast between the dome of St Paul's and the transparent glass spire of the Shard reinforces the cathedral's silhouette."[25]

The tower survived the process. In 2003, the deputy prime minister gave his approval for the Shard. Then the global financial crisis shocked the real estate market, threatening the Shard's survival. A Qatari consortium including the Qatar National Bank bought an 80 percent stake, with the Qatari royal family purchasing two apartments, each the size of an entire floor.

Critics wondered why Londoners were giving away their cherished view for a "safety deposit," the parking of foreign money. They saw it as the largest example of "lights-out London"—the dark and mostly empty homes throughout the city bought by foreign investors. Adding fuel to the fire, the Shard rose beside some of the poorest wards of the city. The tower could raise prices and squeeze Londoners out of the market, potentially pushing a tide of gentrification from the Thames to the south.

But it was already too late! The tower was completed in 2013. Near the Gherkin, in the city's heart, it heralded even more skyscrapers. The historic district of the City of London, with the height restrictions now eased, became prime real estate. This epicenter of finance, bounded on the south by the River Thames and on the north by the old Roman wall, measures only slightly larger than one square mile. Currency dealers in trading rooms there run $2.8 trillion of foreign exchange a day—43 percent of all global capital flows daily.[26]

In 2014, one year after the Shard, two massive skyscrapers were opened in the City. 122 Leadenhall Street has a shape with an iconic lean, to prevent interfering with St Paul's Cathedral's sight line. This noticeable taper, 738 feet tall, along with the building's exterior-facing structure, makes it look like an object used to grate parmesan cheese. Its moniker became the Cheesegrater.

Of all the buildings, the Walkie-Talkie got the most heat. The building, actually called 20 Fenchurch Street, got its nickname for its unusual bulge in the middle, giving it a cartoonish profile on the skyline. "There is perhaps no greater symbol of human idiocy and greed than the Walkie-Talkie," said the writer Alain de Botton, "which has perhaps done more than any building to destroy central London."[27]

His point about destruction rang somewhat true. The building's glassy, concave shape acted like a lens, unintentionally concentrating sunrays and directing them to the ground. It focused the beams so strongly that they melted plastic car parts and burned holes in doormats. Some people even took the effort to demonstrate that the rays scorched the sidewalk hot enough to fry an egg. Editors had a heyday, coming up with new nicknames, including the "Fryscraper."

Architect Rafael Viñoly defended his actions, claiming he had proposed sun shading that would have avoided the hot rays. However, the developer had value-engineered these out of the project. Nevertheless, Viñoly himself had already encountered a similar problem in his design for another concave building, a hotel in Las Vegas known for its "Death Ray." The reflective facade reflected the desert sunrays so intensely on the pool below that it could burn hair.

Even if the Walkie-Talkie may not burn you, it might blow you away. The actual wind impact of the tower on the existing streets in the dense historic City of London differed with the expected wind conditions. This was not foreseen in the environmental assessment of the project's planning application. Several people were nearly blown into the street. It led to yet another moniker, the "Walkie Windy."

This bad situation became even worse when the Walkie-Talkie received the prize for the ugliest building, awarded in the annual "Carbuncle Cup." The jurors denounced the tower for being "a gratuitous glass gargoyle graffitied on the skyline."[28]

In its defense, the project promised Londoners a publicly accessible sky atrium. This unique skyscraper experience would give them a bird's-eye view over a London yet unseen. While the observation atrium was free to the public, there were drawbacks. You needed to book days in advance. When you got there, you had to humble yourself through airport-style security. Upon arrival at the top, the large steel columns that hold up the structure made the space feel claustrophobic, while eating away much of the view.

Besides, this generous gesture to Londoners had an ulterior motive. The tower's position just outside the City's main skyscraper cluster required special permission. City officials preferred packing all the new towers together, rather than spreading them apart, in order to reduce their impact on the skyline. The atrium, initially billed as a garden, helped ensure that the tower received planning approval.

The views from Hampstead Heath and Primrose Hill would never be the same again. Even Prince Charles publicly complained. "Not just one carbuncle, ladies and gentlemen, on the face of a much-loved friend," he said, "but a positive rash of them that will disfigure and disinherit future generations of Londoners."[29]

More critics took aim. "Walking along the Thames has become a game of spotting the latest architectural monster, and remembering its moniker," lamented de Botton. London, he claimed, had lost itself in a "highly misguided search for the tallest and most 'fun' building."

Granted, not all of these "fun" buildings were true skyscrapers or were located in the center of London. The "Armadillo," referring to the oval-shaped, ribboned City Hall, was only a ten-story building. The "Helter Skelter," the 376-foot-tall Orbit Tower, was an observation tower, and the fire-engine-red structure actually looked more like a looping roller coaster. The "Razor," a 486-foot-tall office building with a sharp top pierced by three circular wind turbines reminiscent of an electric shaver, stood in south London.

Elsewhere in the city, dozens of new luxury residential towers spread out without a grand plan. Each borough decides for itself what to approve. The boroughs are eager to develop for revenue, with the tall buildings, like wildfire, "erratically scattered, like needles, across the landscape," de Botton critiqued, "destroying vistas and sightlines at every turn."

Manchester, England's second-largest city, underwent a similar fate. Until 2006, there were only two towers taller than 100 meters (328 feet). Then came ten more, with another twenty in construction.[30]

These towers, many of them luxury housing, stand in stark contrast to the city's homelessness and housing affordability crisis. "The city sold its soul for luxury skyscrapers," the *Guardian* architecture critic Oliver Wainwright wrote. Engels would be shocked to see these swanky towers rise in what he once described as "Hell upon Earth."[31]

Napoleon III would have been equally surprised to see the twenty-first-century skyline of Paris. A stroll along the boulevards of Paris no longer fixates views on the Arc de Triomphe, but also on three new skyscrapers. These towers not only rejected Haussmann's height restrictions; they overcame the assumption that the ground of the central city, weakened by the Catacombs of Paris, was inadequate for high-rises. This elaborate tunnel system dates back to the

Tour Triangle, Paris, Herzog & de Meuron, 2020

Roman Empire. The known hideouts for criminals and a graveyard for more than six million people snakes below the center of the city, where it weakens building foundations.

Paris, the City of Light, is no longer afraid of height. The 600-foot-tall Tour Triangle, designed by two Swiss architects, even dares to vie with the Eiffel Tower for dominance on the skyline. Initially, the project, containing a hotel and office space, was rejected by the Paris city council. When the developer ceded space to childcare and cultural centers, Paris's mayor caved.

The triangular shard is still on the fringes of the historic center. To build within Paris's inner walls, as the new 525-foot-tall Palace of Justice is doing, remains more controversial. Renzo Piano, tasked with the job, designed a ziggurat-shaped tower, four stacked boxes each receding from the one below, just west of Montmartre. Architect Jean Nouvel's site for the Tours Duo is even more conspicuous. These two opposite leaning twin towers, one of them as tall as 590 feet, stand on the banks of the Seine.

Meanwhile, two miles west of central Paris, the La Défense district is the site of seven new skyscrapers. Built with the prospect of beckoning business from London, the tallest of these, Hermitage Plaza, will be another twin tower. Sir Norman Foster, the architect behind London's Gherkin, created the pair with twisting forms tapering outward to the top, with each sibling taller than the Shard.

In London and Paris, so it seems, the tide appears to continue to swing in favor of the high-rise towers. Some of the better designed projects may have the potential of becoming new landmarks for the city, perhaps on par with Big Ben. The Gherkin already stands as a symbol for London, especially since its scene in *Harry Potter and the Half-Blood Prince*.

Now, as most of the public seems to embrace the tall newcomers, old regulations are finding new interpretations, including the view corridors, once obeyed at all costs. "It would be difficult to argue that you could breach the foreground of the view," said the City of London's head of design. "But there's more of a debate to be had when buildings breach the background."

Even the cherished background views from Hampstead Heath may be up for grabs soon. "Many of the Heath's strollers, snoggers and dog walkers," the *Observer* architecture critic noted, "will not give a second glance to this architectural knifing dimly visible through the haze."[32]

Since the city has relaxed its regulations, lesser architectural icons are finding their way. The "Trellis," or 1 Undershaft in the City of London, will rival the Shard in height, and will dwarf the nearby Cheesegrater, Gherkin, and Walkie-Talkie. Built as a plain, long box with exterior bracing, it has little of the formal sophistication of the Gherkin or the idiosyncratic lean of the Cheesegrater.

In this onslaught on the skyline, the City of London's tower cluster now easily outdoes St Paul's, a symbolic shift in power from the church to global finance. The towers will also further cast shadows on the City's network of narrow lanes. When enough Londoners object, the tide may swing back to favor stronger regulation once again.

Until then, without public backing, the critics are left with few options other than taking to their pens. Back in the nineteenth century, author Guy de Maupassant, part of the resistance to the Eiffel Tower among artists and writers, hated the tower so much that he would go and have lunch in the structure's restaurant. It's the only place in Paris, he would say, where he doesn't have to see it. Soon, replicas of the tower were produced and displayed on corners throughout Paris. The public had accepted the tower. In this "inevitable and racking nightmare,"[33] he wrote, "I left Paris, and France, too."

LONDON'S LOW-RISE SKYLINE used to get its few peaks from St Paul's, Big Ben, and Tower Bridge. Today, it has hundreds, with the latest arrivals including the Can of Ham, the Vase, and the Flower Tower. In 2019 alone, 60 new tall buildings were completed—defined as buildings taller than 20 stories—with 525 more tall buildings in the pipeline.[34] This is not to say that everything goes. London's new mayor, Sadiq Khan, blocked the Tulip, a Norman Foster–designed 1,000-foot-tall observation tower with its eponymous shape, citing the protected views of the Tower of London. Protests stopped Renzo Piano's proposal for the borough of Paddington, nicknamed the Paddington Pole, a 72-story glass cylinder that would have towered over the lower historic surroundings.

Now that London is no longer just a low-rise city, city planners are responding with new regulations to protect the public realm. In 2020, the City of London created the nation's first wind microclimatic guidelines. These require developers proposing towers taller than 100 meters to commission sophisticated wind tunnel tests to avoid dangerous downdrafts on the street. This helps keep passersby safe. In 2011, a wind gust caused by a tall tower in Leeds tipped over a truck, killing a pedestrian.

Many more regulations will impact tall buildings. In 2005, the UK's Environmental Protection Act was amended to make light pollution a statutory nuisance. Light pollution has been associated with health detriments, with an impact on our sleep quality, circadian clock, and even hormones. These regulate much of the human body and depend on the cycle of brightness and darkness. External lights also disrupt the migration paths of birds and can cause fatal window clashes. Cities around the world, from London to Chicago, have implemented a "lights out" program. The program encourages tall buildings to keep their lights out at night during migration seasons.

Some regulations will profoundly change how skyscrapers are made. In 2018, the mayor of London declared a climate emergency, committing the city to carbon neutrality by 2050. France also plans to be carbon neutral by 2050. This is not just about adding photovoltaics or light occupancy sensors to buildings. Both the UK and France will require developers to conduct whole life-cycle carbon assessments for buildings. They may put a cap on a project's embodied carbon, the energy necessary to construct the building and its materials. France, the country that invented reinforced concrete, is now emphasizing bio-based materials such as wood, which have a smaller imprint on the environment. The new London Plan requires new building proposals to show plans for building components' disassembly and reuse. This will lead to skyscrapers designed for deconstruction, favoring modular projects and with mechanical connections such as bolts instead of non-removable glues or welds.

Other regulations increasingly encourage tall buildings with a mix of uses, instead of a single use, such as the latest guidelines by London's City of Westminster.[35] Mixing uses in multilevel structures dates back to ancient Rome, with Trajan's Market believed to contain shops, apartments, and possibly Emperor Trajan's administrative offices. In the twentieth century, however, the mass-produced automobile centrifuged and segregated a city's functions: living in suburbs, working in office parks, and shopping in suburban malls. Many of our cities and buildings also reflected the modernist planning ideal of functionalism, where each room, building, or city district was defined by a single purpose. In the 1960s, French philosopher Guy Debord critiqued how modernist city planning was a "technology of separation,"[36] with urban planners of postwar Paris isolating people in the suburban *banlieues*, the social housing projects on the city outskirts. Single-use high-rises are also at risk of becoming dormitory towns of their own. "Stacking people on shelves is a very efficient method of human isolation," said urbanist Jan Gehl.[37] In his novel *High Rise* (1975), British author J. G. Ballard depicts social life in a luxury high-rise building crumbling into violent chaos. "The high-rise was a huge machine designed to serve, not the collective body of tenants, but the individual resident in isolation," writes Ballard. Even conventional office buildings resemble dystopian scenes from Jacques Tati's *Playtime*, where people spend their entire day solitarily, each confined to their own private cubicle.

Perhaps in twentieth-century urban planning we have lost some of the potency of cities' etymological roots—*civitas*, which once meant community. These days, planners, departing from their modernist predecessors, are trying to undo this damage. Instead of zoning districts to be single uses, they are advocating for mixed-use communities, where people can reach their local store, health clinic, school, or office within a short walk. At the same time, the real estate market is discovering the benefits of mixed-use communities as well. In the United States, hundreds of single-use malls have gone out of business. They are being redeveloped as condominiums and townhouses with ground-floor retail and restaurants. With new residential development comes more foot traffic at the mall's leftover stores.

Tall buildings also benefit from mixing uses. For developers, it brings economic diversification, such as revenues from hotels, apartments, and offices in the Shard. For users, it brings convenience and social vibrancy, increasing levels of daytime and nighttime activity. Where twentieth-century supertalls used to be predominantly office towers, the majority of twenty-first-century supertall towers mix uses, such as offices, apartments, and hotels, all under one roof.

In a weird way, boundaries between offices and residences are blurring as well, a throwback to our buildings of the classical age, when we did not separate work from home. Instead of entire office floors dedicated to single tenants, in coworking spaces anyone can rent a single desk, sit on a sofa, plus play Ping-Pong. Where the work floor is starting to look like someone's living room, residential layouts of co-living towers remind us of long-stay hotels. Tenants have small apartments, but with access to shared community space and possibly a business center. Hotel rooms, in exchange, are starting to resemble small offices and lobbies appear like giant-sized living rooms with long tables and people working on laptops.

In projects like this that emphasize common spaces, tenants may be more prone to "collisions," those serendipitous encounters with a potential fruitful outcome, like a perfect idea you could have never thought of alone. Knowledge-based industries benefit from spatial clustering to facilitate face-to-face interaction, where unpredictability and novelty can help stimulate the brain. Researcher Thomas Allen, a former professor at MIT, wrote of the unintended benefits of the university's "infinite corridor," the long hallway that organizes most campus classrooms. Allen was known for the "Allen Curve," finding that collaboration increases as a function of proximity.[38] He observed that in between classes, the crowded corridor became a social condenser of random and serendipitous encounters. These contributed to several inventions. The university eventually chose to plan future buildings around this corridor, to further reap its rewards.

The legendary innovator Steve Jobs, cofounder of Apple, famously planned Pixar's new headquarters with a large atrium at the center. He then located everything from the meeting rooms to the cafeteria, mailboxes, and even the bathrooms beside the atrium. This way, people would be more likely to run into each other, increasing the possibility of serendipitous encounters.

What will this mean for skyscrapers? Instead of segregating people on stacked floors, towers could feature large atria connecting people from different levels. The Zaha Hadid–designed Morpheus Hotel in Macau features a pair of towers interconnected by amorphous skybridges inside free-form atria. They could feature

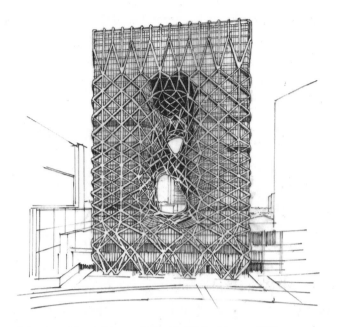

Morpheus Hotel, Macau, Zaha Hadid Architects, 2018

communal gardens, like VIA 57 West in New York, a tetrahedron-shaped high-rise designed by architect Bjarke Ingels, which infuses the skyscraper with a vertical courtyard. Tall buildings could bring people together through sky lobbies, roof decks, and outdoor community spaces—preferably truly public ones.

Still, none of this guarantees diverse encounters, however. The quantity, quality, and affordability of housing is a concern in many European cities.[39] How can we create more diverse skylines? Capping prices may not work, possibly leading to lower housing quality. Building more housing, for instance by incentivizing developers to produce affordable housing, could help drive prices down. This could mean more high-rises for European cities.

With good planning, the next generation of tall buildings does not have to doom us to a "siloed" and segregated existence—from dormitory apartment to office cubicle, reverse, and repeat. Nor should their development diminish the public realm. After all, public space needs to be open and attractive to draw people from their private hideouts. And therefore each new addition to the vertical city requires planners to assess its impact on the horizontal city, and reevaluate the rules that shape skylines.

VIA 57 West, New York City, Bjarke Ingels Group, 2016

SIX

The Competition for Air Rights: New York

Today, New York buzzes with frenetic construction activity, cranes dot the skyline, and scaffolding wraps entire blocks. Skyscraper after skyscraper shatters records. The Empire State Building, once the world's tallest building, doesn't even make it into New York's top six. But the biggest surprise may be the newcomers' figures. Many of the tallest buildings are extremely skinny, almost like needles on the skyline. Unlike the Empire State Building, famed in part for its monumentally stable outline, these new towers seem to defy both gravity and common sense. They have narrow waistlines and often jagged silhouettes. Still, they are inching toward the sky, craning far above the rest of Manhattan.

Like so many other cities, New York is experiencing an unprecedented skyscraper boom. Its protagonist is an unusual and new building type: the "super slenders," the skinniest buildings in the world. These skyscrapers have been rising all over the world. Hong Kong and New York already have dozens. Their skinny silhouette may become a familiar sight from Melbourne to Toronto. Even the tiny mountain town of Vals in Switzerland, known for its magical ratio of a thousand people and a thousand sheep, has proposed a super slender.

Known as pencil towers, these super slenders make the average skyscraper look fat. Most structural engineers consider a building slender if it's at least seven times as tall as it is wide, with a so-called slenderness ratio of 1:7. Others believe a slender building is ten times as tall as it is wide (slenderness ratio 1:10). But it's a subjective standard. Most people know when a tower is slender, just by looking at it.

Even by these standards, New York's latest additions are at the extreme end of the skinny spectrum. 111 West 57, an 84-story skyscraper, is the world's tallest and most expensive condo building. It has the proportions of a pencil: it's a whopping twenty-four times taller than it is wide (slenderness ratio 1:24).

In their extreme slenderness, these towers break the conventional skyscraper design principles. The taller a tower, the thicker the elevator core, the more area dedicated to mechanical spaces, and the more structure necessary to handle the wind forces. But every square inch of shaft or duct is a square inch that cannot be sold. To offset these extra spaces and costs, conventional skyscrapers counterbalance all their lifelines with enough usable space, by thickening the area around the core. Super slenders shatter that economic dogma.

Super slenders have a lot of bone and little meat. In their skinniness, they're worth more. Their smaller floor plates are suited to whole-floor luxury apartments, giving tenants panoramic views from every side, rather than just one side. Since they rise far above existing skyscrapers, they provide more privacy as well.

It should be no surprise of all places, this building species has thrived in Manhattan. From early on, the island had been predisposed to tall buildings. Constrained by water, the borough grew upward instead of outward. Manhattan became vertical. As early as the 1880s, rising land values incentivized developers to build higher. But, as Manhattan grew upward, problems came. With the influx of tall buildings, some streets became dark canyons. The system had to change.

New regulations capped overall vertical growth. Others gave developers a unique ability to trade rights to build higher on a particular site. Even without buying the land below, developers could transfer air rights from multiple blocks away. It was the ultimate commodification: the city allowed developers to buy and sell air. They could now accumulate vast amounts of air rights, even on a tiny site. With enough air rights at their disposal, their buildings could punch a hole in the sky.

New York's tallest skyscrapers used to be office towers, symbols of corporate power. Today, they are luxury residential towers containing penthouses, the playgrounds of the world's billionaires, with a single apartment going for a hundred million dollars. To some, the super slender apartment represents a status symbol, outshining a Maserati car or a limited-edition Gucci handbag.

Yet this comes at a public cost. The towers on Billionaire's Row, the block of super slenders just below Central Park, are casting long shadows deep into New York's green heart. Most of the towers sit largely empty—if tenants spend less than half a year in the city, they can avoid paying hefty New York City taxes.

Meanwhile, the super slender tenants have problems of their own. They are finding life in ultra-skinny buildings, which at times sway in the wind like reeds, has pitfalls.

"BUY LAND," Mark Twain once said about real estate. "They're not making it anymore." While land may now rarely be made, especially since sea-level rise is threatening land reclamations, it is really not that scarce. You could fit the entire population of the United States into the state of Texas, with more than an acre per household.

What drives up the price is less the availability of land, and more our collective desire. We don't like every piece of land equally. Some 1.6 million people choose to live in Manhattan, an area of only 23 square miles. This makes for a population density of more than 70,000 people per square mile, among the highest in the world.

With this limited supply of desirable land, and backed by increasing demand from urban population growth, prices have risen swiftly. Back in the seventeenth century, a director of the Dutch North American colony of New Netherland gave the Lenape Native Americans a scant $24 worth of trinkets for Manhattan Island, an amount worth a little over $1,000 in current dollars. Today, an acre of central New York costs more than $123 million.[1]

Some governments deliberately reduce the supply of buildable land to generate more revenue. In Hong Kong, the government's restrictions inflate land prices, which the city depends on for a large portion of its tax revenue. In contrast, the island of Manhattan, already fully built, has little control over its land supply. The city manages to control buildable areas through other means—by controlling how high people can build upward. It regulates the air.

In most cities, including New York, zoning and building regulations control both how to build and how much to build. The first building code recorded in the United States dates back to 1625, when the city then known as New Amsterdam regulated roof coverings to protect from chimney sparks. The most influential instrument to regulate New York's growth was the Commissioners' Plan of 1811. It overlaid a grid on top of Manhattan. In almost complete disregard of the early settlements and natural topography, the grid turned everything into rectangles of 264 by 900 feet.

This rigid grid projection onto the pastoral spaces of mid and upper Manhattan neglected existing buildings, streams, vegetation, and even hillsides—all for a bigger purpose. The gridiron represented a republican ideal. The equally spaced layout avoided the utopian urban schemes of Europe, its uniformity emphasizing that governments should not act to privilege one over the other.[2] It had other benefits as well. "Straight-sided and right-angled houses are the most cheap to build and the most convenient to live in," the commissioners of New York noted.

Unlike London's organic plan, Manhattan's gridiron with its equal-sized lots was geared toward real estate speculation. It was as if the commissioners were saying this town was not going to be reigned over by autocrats but by the market.

Within the private lots of the blocks, those inflicted with a lack of light as a result of neighboring projects would have no recourse. Up until 1838, US courts applied British Ancient Lights. But then the court had to rule about a property owner whose new store would prevent almost all light to the adjacent dwellings. "It may do well enough in England," the court stated about the law, siding with the store owner. "But it cannot apply in the growing cities and villages of this country, without working the most mischievous consequences."[3]

Without the law, American cities could densify with few restraints. In a weird twist of irony, this was a stark departure from the nation's founders who had feared dense cities. "The mobs of great cities add just so much to support of pure government as sores do to the strength of the human body," Thomas Jefferson once declared.[4] However, in an age of US economic expansion, the right to protect property had superseded the desire to avoid crowds.

The regulatory stage for tall buildings was set. In the nineteenth

century, the country's urban population exploded from 6 to 40 percent.[5] With few restrictions to build, the world would soon see its
first "skyscrapers," named after a triangular sail set above a ship's
skysail. The name stuck.

Skyscrapers appeared first in Chicago, despite the city's unsuitable soil—or its "slimy ooze," as the *New York Times* noted in 1891,
asking, "Who shall restrain the great layer of jelly in Chicago's cake?"[6]
In 1893, Chicago put a height limit of 130 feet on buildings, out of fear
for safety and a lack of light on the streets. Without such qualms,
New York was destined to take Chicago's place. Manhattan, with its
solid bedrock, needn't worry whether the soil would keep the tall
towers aloft. The bigger problem was that the sky would be gone.

Manhattan's living conditions became dreadful. By the late
nineteenth century, most of the grid was filled in with development,
with 60 percent of the city living in substandard tenement dwellings.[7] In 1894, the Tenement House Commissioners wrote: "[T]hou-
sands of people are living in the smallest place in which it is possible
for a human being to live—crowded together in dark, ill-ventilated
rooms, in many of which the sunlight never enters and in most of
which fresh air is unknown. They are centres of disease, poverty,
vice, and crime, where it is a marvel, not that some children grow up
to be thieves, drunkards and prostitutes, but that so many should
ever grow up to be decent and self-respecting."[8]

New Yorkers also worried about increasingly larger buildings.
The writer Henry James described skyscrapers as the "monsters of
the mere market," overshadowing churches, which were "merci-
lessly deprived of their visibility."[9] A landscape architect saw the
towers as "a revolt against the laws of Nature ... piling humanity up
in heaps like bees or ants, absorbing and disgorging them twice a

day until the streets become too narrow for the traffic and the sewers too small for the drainage they have to carry away."[10]

Soon enough, disaster struck. In 1911, a fire broke out at the Triangle Shirtwaist Factory, a ten-story building. The owners, fearing theft, had blocked the only accessible fire exit. The fire became the deadliest industrial disaster in New York, killing 146 garment workers, including girls as young as fourteen.

In response to this tragic fire, the city created its first high-rise building code. More laws followed requiring automatic sprinklers, fireproofing materials, and fire extinguishers and alarms. The city also considered new zoning regulations to protect the public from what are known as "negative externalities" of new buildings, such as a lack of daylight. But the real estate industry's objection put these to a halt.

In 1914, however, an outrageous new Manhattan skyscraper took things too far. The Equitable Building, a 40-story monolith, filled an entire downtown city block. It dumped thousands of people on a narrow band of pavement and put much of the neighborhood in its shadow. For its blatant disregard of its context, it was called a "monstrous parasite on the veins and arteries of New York."[11] Property owners filed for a reduction of their property valuation, claiming the shadow had robbed them of significant rental incomes.

A massive zoning overhaul could no longer be stopped. The 1916 Zoning Resolution, the first of its kind in the nation, would completely alter the city skyline for decades to come. A century earlier, the Commissioners' Plan had shaped New York horizontally, by laying out its streets. Now the 1916 plan shaped the city vertically, by putting a cap on skyscrapers in an ingenious way.

The city wanted to avoid the "canyon effect," when tall buildings flank both sides of the streets and leave only a thin strip of light in the sky. The city did not want to impose Ancient Lights, partially to avoid the complicated negotiation between owner and judge that this law required. Instead, the city chose another regulatory mechanism to guarantee light: universal setbacks. They could be transparent, handled by administrators, and wouldn't be as subject to corruption. It turned out that they resolved the issues just as well.

The city's challenge was to design setback regulations so new development would still allow for air and light to get to streets. Planners adopted a method advocated by Boston architect William Atkinson. He argued the shape of buildings mattered. He deemed simply capping a building by a fixed horizontal plane of arbitrary height "not scientific," since the rear portions of buildings don't block daylight on the street as much as the front portions, which face the sidewalk. "The rear portions may well be allowed to rise to a greater height," he wrote.[12]

Instead of a horizontal plane, Atkinson proposed a slanting sight line to cap the building's height. This imaginary line would rise from the ground level of the street diagonally into the sky above the building lot. The further from the street, the higher the line rises. Buildings, not allowed to penetrate this line, were forced to gradually step back.

The city implemented these slanting lines as so-called sky exposure planes, imaginary sloping planes. They specified different slopes for each area. In areas in which the city wanted more density, the line was allowed a steeper slope. The densest area was lower Manhattan, deemed a "2.5 times district." This meant that a line beginning from the center of the street could rise up to two and a half times the width of the street at the front edge of a building. For instance, the line going from the center of a 100-foot-wide street would rise to a 250-foot-tall building edge. From there, buildings would have to set back 10 feet for every 50 feet rise, along the five-to-one ratio of the line. Without these sky planes, New York streets would be in constant shadows.

The architect and artist Hugh Ferriss first visualized what New York would become, later published in his 1929 book *The Metropolis of Tomorrow*. He understood that Manhattan's builders would simply build whatever was the maximum allowable on the site. While theoretically the diagonal sky planes, applied to rectangular city blocks, would lead to pyramid forms, diagonal walls are not practical to build. Most developers chose tiered skyscrapers, gradually tapering shapes whose taller parts become smaller toward the top. This led to buildings that looked like a wedding cake.

Ferriss's illustrations revealed the pure expressions of this new code, combined with the economics of building. His dark charcoal drawings delineated a dense forest of tiered, pyramid-shaped buildings. He often presented the buildings at night, adding light beams, fog, and rooftop landing pads. His drawings would inspire Gotham City in *Batman* movies.

Truth is stranger than fiction. The Roaring Twenties soon would see Ferriss's drawings materialized, each block with its own innovative take. By 1925, New York had overtaken London as the world's most populous city. Now a true "race into the sky" emerged, with the Chrysler Building and the Empire State Building—both tiered, wedding-cake-style skyscrapers—at its center.

With the new rules in place, architectural shape became an economic equation. Developers would figure out the maximum they could profitably build on a site, and this would become the form. It was "form follows finance," as the architectural historian Carol

Chrysler Building, New York
City, William Van Alen, 1930

Willis has noted.[13] Architects were relegated to decorating the given
forms with geometric patterns.

In ancient Egypt, the skyline was an expression of the rule of the
pharaoh. Now the skyline of Manhattan followed the rule of the real
estate market. Builders built the maximum the regulations allowed.
Skyscrapers there became the purest expression of the zoning codes.

It so happened that New York's wedding-cake towers fit well
with the vertical articulation of Art Deco. Architects would emphasize their setbacks. This led to the Ziggurat style, a stepped pattern
of Art Deco skyscrapers, as in the crown of the Chrysler Building,
built in 1930. Architect William Van Alen created a peak of seven
curvilinear surfaces gradually stepping back into a point. Each is
decorated with radiating sunburst patterns and reflective Nirosta

steel. Today, the crown's shiny patina signals the exuberant energy of New York like no other building and is a paragon of Art Deco. However, at the time, some New Yorkers were tough on this shiny newcomer on the skyline. A writer in *The New Yorker* dismissed it as "merely advertising."[14]

Within a year, the Empire State Building would become the tallest building in the world, at 1,250 feet taking the crown away from the Chrysler Building. While its top was not as decorated, it was to become graced with an iconic image in *King Kong*, the 1933 film. In it a giant ape in his rampage climbed the Empire State Building, one hand clutching Fay Wray while the other swung at fighter planes.

Empire State Building, New York City, Shreve, Lamb & Harmon, 1930

Although buildings this tall were possible by regulations, they did not automatically make economic sense. According to the economist Jason M. Barr, each building has an "economic height," the height at which a developer can get the most profit out of a piece of land. After this height, given the higher construction costs for taller buildings, profit diminishes. To have maximized economic profit, the Empire State Building would have had to have been 54 stories lower.[15] Barr called buildings like this subject to the "status effect"— a developer willing to trade in maximum profit for the sole purpose of status. While real estate surely has pumped up New York's skyline, without anyone willing to stroke their ego the city's skyline would have been quite a bit lower.

There were other irrational aspects in this early phase of New York's skyscrapers as well. Before a mature real estate market would push them to become standardized products, skyscrapers contained a kaleidoscope of activities, all under one roof. The architect Rem Koolhaas celebrated the skyscrapers of this age for their "culture of congestion," with their seemingly absurd juxtapositions of different uses. He admired buildings like the Downtown Athletic Club on West Street, a 518-foot-tall building that included an interior golf course. One floor featured a boxing club, a locker room, and an oyster bar. It was almost the set of a surrealist dream. "Eating oysters with boxing gloves, naked, on the 9th floor,"[16] Koolhaas wrote in *Delirious New York*. Skyscrapers were social condensers, with seemingly random uses bringing people together in entirely new ways. They embodied the great promise of cities.

The project most representing the New York skyscraper's potential for greatness was arguably Rockefeller Center. Touted as a city within a city when it was first commissioned by John D. Rockefeller in the 1930s, it married commercial and cultural functions into a complex of nineteen buildings, each an example of the wedding-cake style. The project even included a public plaza in front of the tallest skyscraper, which in the winter holds an ice-skating rink. Shorter buildings surround the center tower, each lavishly decorated with art, one of them housing Radio City Music Hall. An underground concourse connects a shopping complex and the subway system. A common saying held at Rockefeller Center: "You could do anything you wanted except sleep (no hotels), pray (no churches), or not pay rent to [Rockefeller] Junior."[17]

By the 1930s, Manhattan featured hundreds of buildings that could be called "skyscrapers." The city even had a few early super slenders. These towers rose as a result of an exemption to the sky planes. A quarter of the property area, located in the middle of the lot, could go up past the planes. The Pierre, a luxury hotel completed in 1930, only had a narrow waist, yet rose to 525 feet. It was topped with an oxidized-copper mansard roof modeled after the Royal Chapel of the Palace of Versailles. Its tall shape was justified by its Central Park location, with hotel guests willing to pay up for a view.

The Great Depression and World War II put a halt to more of these skyscrapers. By midcentury, city planners were questioning the principles of the 1916 Zoning Resolution. They realized if New York had been fully built according to the resolution, its population would have far exceeded the city's realistic capacity, with up to 55 million people.[18] Meanwhile, modernist ideas about urban planning had crept into their minds, including the "tower-in-the-park." This is a minimalist, boxy tower standing in a sea of open space, an idea propagated by the modernist architect Le Corbusier.

The new 1961 Zoning Resolution regulated the overall building volume rather than the form. Instead of a particular envelope determined by sight lines, it focused on a mechanism called "floor area ratio" regulations, a maximum amount of bulk an owner could build. For each block, planners would establish a density quota. Still, this did not prevent tall buildings even in low-density areas. If the owner's lot was large enough, all the available buildable area could be stacked on a small portion of the site, allowing a skyscraper next to low-rise buildings. The new regulations thus allowed developers to neglect the scale of the local context.

Less constrained by diagonal sight lines, builders could build straight towers and make them as large as the maximum floor area allowed, as long as they built in the central portion of the site. The new zoning resolution led to a new breed of skyscrapers that departed from the Art Deco wedding-cake style. In its place came International Style glass boxes.

Besides these less exciting geometries, New York skyscrapers also lost their multiuse vibrancy. The real estate market had matured, and towers became either residential or office, each with their own respective standard floor heights and floor size. Office towers became "core and shell," with empty interiors left to be fitted out by individual tenants. With building codes regulating the elevators and egress stairs on the inside of towers, architects spent most of their creative energy designing the curtain wall exterior. In 1973, these regulations spawned the imposing Twin Towers of the World Trade Center, briefly the tallest buildings in the world. For their plain extrusions, urban critic Lewis Mumford called them "just glass-and-metal filing cabinets."[19]

In addition to the modernist towers, the city's planners had been captivated with a few new private plazas. The minimalist plaza lined with reflecting pools adjacent to the Seagram Building, a burnished copper building designed by Mies van der Rohe, buzzed with life. Planners realized the city could benefit from more open space.

The 1961 Zoning Resolution included an incentive that would allow builders to build more volume, if they were to provide an adjacent public space. For every one square foot of so-called bonus plaza they provided, builders could build ten square feet more. For developers, the deal was too good to pass up. By 1973, some 1.1 million square feet of new urban open space had been built in New York City by the private sector, more than in all the other cities in the United States combined.[20] But while the city succeeded in leveraging private development to increase open space, most of these plazas were little used.

"The city was being had," asserted William H. Whyte, journalist and urbanist. The New York City Planning Commission assigned Whyte to investigate the problem. Along with a group of researchers, armed with camera and notebook, he started to document the street life of these urban spaces. He noted that the plazas had little seating. When they did have places to sit, such as ledges, they were often outfitted with spikes. Developers built open spaces for their visual appeal, not for public use. "For the millions of dollars of extra space it was handing out to builders, it had every right to demand better plazas in return,"[21] Whyte concluded. Whyte's advocacy led to a 1975 amendment, forcing plazas to be "amenable" to the public, eventually leading to better open space with more seating.

Where the public plaza provision of the 1961 Zoning Resolution would provide New Yorkers with more open space on the ground, another change would fundamentally alter their skyline. The revamping of the city zoning regulations would make air rights a salable commodity. It allowed the transfer of air rights between properties that share a minimum of ten feet of their lot line, through a mechanism known as "Transferable Development Rights." When investors buy the air rights of an adjacent property, they can then buy the air rights of the lots adjacent to their recently purchased property. And so the game goes on. It would lead to air rights being combined over multiple blocks, snaking from one side of the neighborhood to the next. As a result, investors could assemble an almost infinite amount of air rights, and build to unprecedented heights.

In New York, air is invisible land. In rare cases, the air above the land can even be worth more than the actual land below. Open sky above a shorter building is a tall tower yet to come, either where you see it or many blocks away. Call it "unreal estate."

Air rights have existed since 1797, with a British court decision in the Colony of New York over two stolen barrels of herrings. They had been buried 15 feet deep under a property owner's yard. The decision relieved those who feared trespass from below their properties, or from above.[22] The court decided that anything below or above the soil belongs to the property owner. This was based on the Latin phrase *Cuius est solum, eius est usque ad coelum et ad inferos*, or "For whoever owns the soil, it is theirs up to Heaven and down to Hell."

This definition was refined after the first hot air balloon flight in 1783, when it became clear that *ad coelum* ("up to Heaven") would make air travel practically impossible. In the early twentieth century, a court redefined the space above the soil as "within the range of actual occupation."[23]

Air rights were first sold in New York in 1908 during the construction of Grand Central Terminal. The railroads could help fund the terminal's construction by selling the air rights above the terminal's tracks and platforms. This spawned housing and office towers over nine blocks. But the 1961 Zoning Resolution vastly expanded the market for air rights, since developers could more easily add them to their lot. The air rights of landmarks were especially easy to accumulate. City planners compensated historic buildings for their potential losses as a result of their landmark status, which restricts expansions. Owners of landmark buildings could now sell their air rights to sites across the street, or even several blocks away.

In 1998, planners granted Broadway theaters even more flexibility. Low-rise theater owners could transfer their air rights to anywhere in the theater district, no matter the location. The city government decided that Broadway should be allowed this right in order to support the industry. It would lead to a quintessential New York scene that would make some preservationists shiver, with massive skyscrapers dwarfing tiny landmarks—and ironically protecting them at the same time, by providing them with cash for their air rights.

One of the first to take advantage of the more easily transferable air rights was Donald Trump. In the 1970s, he paid $5 million for the air rights above a landmarked building on Fifth Avenue. Together with other air rights and bonuses, this allowed him to inflate a 20-story building into the 58-story Trump Tower.[24] He then did it again in 2001, with the Trump World Tower at 845 United Nations Plaza. He transferred air rights from several properties and created what was briefly the tallest residential skyscraper in the world at 861 feet. Neighbors protested that the 72-story tower would dwarf even the nearby United Nations Headquarters. But, since air rights transactions happen between property owners and developers, the towers were built "as of right." Unlike in London, where tall towers need to go through a public approval process, most of New York's skyscrapers leave ordinary citizens without having a say.

But in 2001 disaster struck. Terrorists crashed two airplanes into the World Trade Center towers. The Twin Towers, which had been a symbol of the world's economic system, became a symbol of tragedy. The attacks raised difficult questions about the nature of tall buildings. "If there are to be new rules for the new warfare," wrote *Wired*, "one of the first is surely this: Density kills."[25]

With New York's World Trade Center towers left in massive piles of rubble, and the iconic pair of skyscrapers missing from the skyline, the city staged a comeback. Not since the Roaring Twenties had New York seen such construction activity. The New York skyscraper returned to the forefront of the city's public debate, with the super slender center stage.

TRUMP MAY HAVE set the stage with his Trump World Tower. But that was nothing compared to what the former diamond dealer Gary Barnett did. Barnett worked quietly for sixteen years to piece together the land and air rights for One57, planning a 1,004-foot-tall ultra-luxurious skyscraper on West 57th Street, just two blocks south of Central Park. Air rights transfers had made Central Park views possible on small tracts of land that previously had none.

Barnett bought the air rights of a string of mid-rises on the block. These rights were worthless to some, like co-op buildings having no plans to redevelop their parcel on their own. They were happy to sell, but in some cases struck a Faustian bargain. By the time Barnett's building was built, it ruined many of its neighbors' views. "Now One57 blocks my view of Carnegie Hall," one of the residents said.[26]

With a slenderness ratio of 1:23, the project was going to be the world's slenderest tower. Early on, its pencil-thin profile was tested by a hurricane. In 2012, Superstorm Sandy devastated coastal New York, leading to damage throughout the city. Lower Manhattan suffered a multi-day blackout of almost every building. Tens of thousands of people living in high-rises lost elevator service, as well as water and toilet refills.

The hurricane hit One57 while it was under construction. The crane's top crumpled like toothpicks, steel poles falling from the sky. For several days afterward, New Yorkers and tourists saw the upper arm dangle far above 57th Street. "You Americans have put on a good show for us," a visiting Brit said.[27]

Unfortunately, the design of the French architect Christian de Portzamparc suffered difficulties of its own. The architect had hoped to endow the super slender's facade with pixelated blue shades to be reminiscent of a cascading waterfall, naming it a "Klimt effect" in reference to the Viennese painter. However, upon construction, little of that vision remained. The curtain wall, with its random pattern of blue shades, more closely resembled the dizzying mosaics of a swimming pool.

Only a year after its completion in 2015, One57 was outdone by 432 Park Avenue. At 1,397 feet tall, it took the crown for the world's tallest residential building, although it was not as skinny, with a slenderness ratio of 1:15. In his design for the tower, architect Rafael Viñoly drew his inspiration from a trash can. In all fairness, he chose a not your run-of-the-mill waste bin, but a 1905 Austrian garbage can designed by avant-garde architect Josef Hoffmann. Viñoly elegantly worked the minimalist grid-like pattern of the wastebasket into the tower, providing a benignly indifferent counterpoint to the garish facade of One57.

At this extreme height and with its super slender profile, 432 Park Avenue even tempts regularly occurring winds. To reduce the wind load, engineers created five double-height openings with mechanical spaces, spaced every twelve floors, so currents could

432 Park Avenue, New York
City, Rafael Viñoly, 2015

blow through. Toward the top, two tuned mass dampers weighing a total of 1,370 tons further prevent the building from swaying too much. This solution helped counteract the oscillation of this skinny tower, which could make its residents motion sick—one of the unique problems reserved for the world's wealthiest.

The many mechanical areas throughout the tower helped reduce the ductwork size, and the tower's voids and tuned mass dampers made the structure less bulky. Together, these mechanical and structural spaces still occupied a quarter of the tower, reducing the efficiency of the building. Nevertheless, the developer may welcome this limitation. Since mechanical space does not count toward the total floor size, developers can build as much of it as they want. By having taller floors dedicated strictly to mechanical areas, often more than three times the height of the average floor, they can build a taller tower, with better views from upper stories, and charge more.

Meanwhile, for his next super slender, the Central Park Tower on West 57th Street, Barnett took another decade to assemble air rights. The Central Park Tower exceeded 432 Park Avenue and retook the title of the world's tallest residential building, and with a slenderness ratio of 1:14. Barnett, who tried to squeeze out as much square footage from the site as possible, had purchased the air rights above the school of the Art Students League of New York. As a result, Adrian Smith, the architect of the Burj Khalifa, faced a difficult challenge. He needed to cantilever this supertall above a third of the lot of the landmark building, a delicate French Renaissance building from the 1890s. It made for an ominous scene, likened by architecture critic Michael Kimmelman in the *New York Times* to "a giant with one foot raised, poised to squash a poodle."[28]

To ensure that the views from his tower's apartments would stay unobstructed, Barnett went so far as to purchase the lease for a building garage in order to stall a neighboring super slender. He then refused to let go of the lease, delaying his competitor's project for more than seven years. Finally, Barnett sold the development rights for the handsome sum of $194 million. This led to 220 Central Park South (slenderness ratio 1:18), a 70-story building designed by Robert A. M. Stern. Stern provided a welcome departure from the glass boxes with a limestone-clad skin. Upon completion, this tower then made an even bigger sale. The tower's penthouse sold for $238 million, the most expensive home ever sold in the United States.

By now, the press called the area of skinny supertalls south of Central Park "Billionaire's Row." The public debate had been turning against the super slenders craning over Central Park. They frustrated neighbors who complained about the long slivers of darkness moving past the park.

"The feeding frenzy of real estate developers must not be allowed to damage the jewel of New York City," a Landmark West representative said, a nonprofit that aims to protect the city's cultural landmarks.[29] The Municipal Art Society commissioned a report to investigate the towers' long shadows. Tall, skinny buildings have a much thinner and longer shadow than a short and stubby one. While the pencil tower's shadows go into as far as a third of the park, since they are so thin they will pass faster, roughly at 3.5 feet per minute on the extreme end.[30] Still, to a person lying in the sun in Central Park's Sheep Meadow, one super slender will cast a shadow on you for a good twenty minutes.

Upon completion of the report, the arts society condemned the pencil towers, if only for their lack of environmental review or community input. They objected to an "Accidental Skyline," with new supertall towers beyond the wildest dreams of the authors of the 1961 Zoning Resolution.

The debate would leave its mark on future towers. The MoMA expansion tower, or 53W53, was planned on a site formerly owned by the Museum of Modern Art. The tower incorporated three floors of new galleries for the museum. The developer commissioned architect Jean Nouvel to design a muscular, exoskeleton tower with structural beams on the outside forming a large shard. The city chopped 200 feet off the top. Upon redesign, Nouvel's dark shard featured a more striking, asymmetrical peak with jagged edges. It would become a remarkable addition to the skyline—but now at "only" 1,050 feet tall, with a reduced slenderness ratio of 1:12. Under heightened scrutiny, and with each of the super slenders trying to

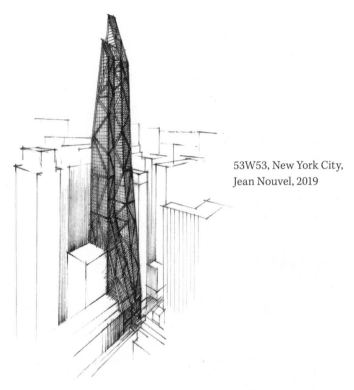

53W53, New York City,
Jean Nouvel, 2019

outdo the other, they gradually became more sophisticated, each
new one learning from the previous one. The slenderness race cul-
minated with 111 West 57th Street, otherwise known as the Steinway Tower. Completed in 2021, the 1,428-foot-tall skyscraper is New
York's third-tallest. Yet it towers on only a 60-foot-wide base, with
a record-breaking slenderness ratio of 1:24. The tower's top slowly
steps back until it disappears into the sky, a throwback to the "wedding cake," with the seeming thickness of a razor blade. The tower
also reflects light more subtly than Midtown's many boxes of glass
and steel, thanks to its glazed terra-cotta tiles, a nod to the city's
early skyscrapers.

While the Steinway Tower shows the potential of the New York
"accidental" skyscraper, it also highlights its pitfalls. 111 West 57th

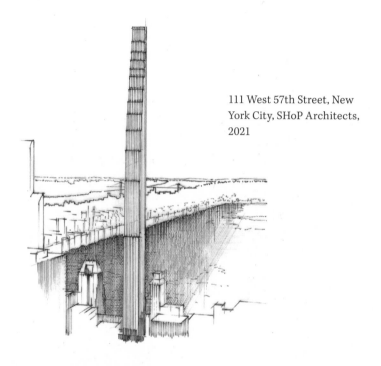

111 West 57th Street, New York City, SHoP Architects, 2021

Street incorporates the Steinway Hall, the piano showroom that once hosted duo performances of Vladimir Horowitz and Sergei Rachmaninoff—now turned into a luxury health club. One critic described the super slender phenomenon as the "plutocratization of the midtown skyline."[31] The towers are the playgrounds of ultra-high-net-worth individuals, who make up only 0.003 percent of the world population, yet hold 13 percent of the world's wealth.[32] The super slender penthouse gives them a tangible asset as an ultimate status symbol. In the process, they are making an indelible mark on the Manhattan skyline.

Extreme wealth is not just pushing New York's skyscrapers upward. From 1930 to 2000, 86 percent of the world's tallest towers were office buildings. In 2020, only 36 percent of the world's tallest towers were offices.[33] More and more, luxury residential buildings are determining the look of our biggest cities. We may expect the uber-wealthy to rise above the skyline in Dubai, where the ruler's thumbprint will be imprinted on the oval-shaped base of Burj Jumeirah, the city's latest proposed supertall—but not in New York.

While air rights transfers fueled the super slenders in Billionaire's Row, endangering Central Park, elsewhere in the city they were arranged to protect a park. The city had enabled the transfer of air rights around the High Line, a park planned on an abandoned elevated freight line in Chelsea, to nearby land parcels. The air rights transfers ensured more light, air, and views in the park, as well as even helping to pay for it.

The High Line linear park has become one of New York's premier tourist attractions. In 2019, it drew more than 8 million visitors, more than the Statue of Liberty. In the meantime, West Chelsea's mid-rise skyline of industrial warehouses became intermixed with contemporary high-rises designed by celebrity architects, with condominium towers by Jean Nouvel, Bjarke Ingels, and Zaha Hadid. The park's designers, New York architects Diller Scofidio + Renfro and landscape designer James Corner, deliberately designed the park to guide a pedestrian's gaze to some quintessential New York sights. Unlike Central Park's bucolic views, the elevated High Line gives tourists a novel way to experience the city's street life, including panoramas of skyscrapers and yellow cabs.

Its success has inspired similar projects worldwide, establishing the so-called High Line Effect.[34] As an architect working on large-scale projects in the years after the High Line's opening, I experienced the project's global appeal. Even in cities lacking any elevated freight lines or air rights transfer scheme, it seemed like there was always someone in the room who wanted to add a High Line.

However, the project's thrill faded when the rapid gentrification of the surrounding neighborhood became apparent. All the
new towers were luxury housing, with a typical apartment selling
for millions of dollars. The air rights transfer scheme did not force
developers to create affordable housing. It had accelerated gentrification, displacing lower-income residents. Even New York's iconic
hot dog vendors couldn't compete with the better-capitalized vendors in applying for vending spots along the park.[35]

The legacy of the High Line is even more at risk at its final stretch,
Hudson Yards, where the selling of air rights created a luxury high-
rise cluster of new magnitude to almost unanimous criticism. In
2010, developers acquired the air rights above the thirty tracks of
the Long Island Rail Road railyard, with plans to build the largest
private real estate development in US history. The entire complex,
including several supertall towers and a mall, is built on a platform
spanning the yard. The whole project exudes exclusivity, with little
relationship to the rest of the neighborhood. Unlike Rockefeller Center, which was elegantly woven into the Manhattan grid, Hudson
Yards is a "superblock." Reluctant to connect to the surrounding
streets, it presents neighbors with mostly blank walls with service
entrances and elevator lobbies.

The first phase of the project opened in 2019 to harsh reviews.
The *New York Times* architecture critic Michael Kimmelman called
it a "gated community." The publicly accessible plaza in the center
of the project, provided by the developer, lacks a sense of place. The
shadows of the scattered bulky towers, the monotony of blue reflective glass, and the car drop-off to the mall all make for an open space
that reminds one, as Kimmelman noted, of "a version of a 1950s
towers-in-the-park housing complex, except designed by big-name
architects."[36] Oliver Wainwright, the *Guardian* architecture critic,

had an even harsher verdict. The headline on his review: "Horror on the Hudson."[37]

Granted, the project had some unique challenges to overcome. Built above the tracks, the plaza was elevated from the surrounding street level. To allow trees to grow on the shallow base, engineers invented a 4-foot-deep "soil sandwich" of sand, gravel, and concrete slab. They even created an underground water reservoir with a one-story underground ventilation system powered by fans usually found in jet engines. These keep the plants cool from the heat of the yard.

If it were up to the developer, the project would have been even more segregated. The developer proposed building a 720-foot-long concrete wall facing the High Line. Gale Brewer, the Manhattan Borough president, managed to stop the barricade.[38] "You don't trust anybody in this world," she said.

Unfortunately, the rich have always tried to remove themselves from the hoi polloi, a tendency urban planners aim to course-correct. In the nineteenth century, many elites departed the congested city for the leafy suburbs. In the same period, New York built Central Park to give urbanites access to greenery as well. In the 1970s, the city aspired to improve social intermingling through inclusionary zoning, giving developers bonus floor area if they included affordable housing in their projects. Developers then built the units but created "poor doors," avoiding full-price tenants bumping into below-market renters. Seeing this, the city responded with a provision forcing developers to create the same common entrances to the affordable units.

Today, developers are trying to take advantage of weird loopholes in the zoning code, which has become so labyrinthine that it is easier to outsmart. Back in 1916, the code was only twelve pages. By 2016, it had exploded to 1,300 pages. It is filled to the brim with special regulations, ranging from the prescribed geometric shape of a bonus plaza to the amount of electronic signage required in Times Square.

As an architect, I frequently find myself lost in the complexity of New York City zoning. Sifting through its hundreds of pages, myriad definitions, and many appendixes can be a bewildering experience. Besides complex legal language and some abstruse provisions, the rules keep changing. Even if you fully understand the codes, there is so much ambiguity that you will still likely need to rely on the interpretation of a zoning official for a final verdict. With each interpretation, a developer armed with the right land-use attorney and an encyclopedic knowledge of prior rulings can build their case.

One developer commissioned Rafael Viñoly—the same architect whose curvy London tower focused sunshine so intensely that it melted car parts—to design a 32-story residential building. This would normally lead to a tower of about 360 feet in height. Nevertheless, the proposed project would reach 510 feet in height, thanks to a 150-foot-tall gaping hole at the bottom. With a tower that is empty for the first fifteen stories, the penthouses can be built higher with better views at the top. This led to the nickname "condo on stilts." A critic described the project as "a *Jetsons*-esque podium to boost its upper levels high into the neighborhood skyline."[39] The developer cunningly classified this void as mechanical space. This way it does not count toward the maximum area that can be built. If all skyscrapers in New York exploited this mechanical-void loophole, the lower 150 feet of the entire skyline would just be pure air.

City planners plotted new rules to stop this gap. The Department of Buildings halted the stilted tower for safety concerns. New regulations proposed to count voids as floor space when they are taller than 25 feet.

Public plazas, vibrant streets, green parks, affordable housing, and efficient subways are the "great equalizers," providing a counterbalance to the tendency of wealth to concentrate. There are more equalizing methods underway, from closing the loop of the mechanical void to capping the outsized carbon emissions that supertall towers would generate.

New York may have put a halt to stilted skyscrapers. But more zoning loopholes will no doubt be found. The Municipal Art Society proposes to revisit the 1961 Zoning Resolution. It is calculated that 3.7 billion square feet of development rights still exist in the city, enough to build 1,300 Empire State Buildings.[40] They argue for a public review for any building exploiting air rights transfers, and a cap on the maximum air rights that can be purchased.

Nevertheless, in New York there is a counterintuitive benefit to tall towers. They protect other parts of the city, allowing them to stay low. Since the overall allowable density has a limit, the tall buildings will not add more density to the city; they will just transfer it from one site to another. Super slenders balance the skyline. Aesthetically speaking, their thin peaks add big crescendos, allowing for diminuendos elsewhere, avoiding the otherwise monotonous solid wall of equally high-pitched buildings.

In addition, where tall buildings used to be the bread and butter of large corporate architecture firms, the supertall skyscrapers of today are increasingly designed by celebrity architects. These architects bring novel designs, like 56 Leonard Street, completed in 2017. While it has a "modest" slenderness ratio of 1:10.5 and a height of 821 feet, the architects Herzog & de Meuron pixelated the tower with "villas in the sky" glass boxes, leading to its nickname, the Jenga Building.

Super slenders also have an aesthetic advantage over office towers. Unlike an office building, where wide floor plates are desired, a luxury residential tower can be as narrow as a single apartment. Slenderness makes for a much more dramatic peak on the skyline.

The new World Trade Center, for instance, opened in 2014 at the symbolic height of 1776 feet, the year of the Declaration of Independence. It has a slenderness ratio less than 1:7, because of its wide office floors. Almost a quarter of the tower's height is comprised of a 408-foot-tall antenna. An antenna does not contribute to the slenderness or height of a building, classified as "vanity height." Nevertheless, an expert panel decided that One WTC, given the gravitas of events, could include the spire in its height measurement. To the frustration of Chicago, the Willis Tower has been dethroned as the nation's tallest. ("If it looks like an antenna, acts like an antenna, then guess what? It is an antenna," Chicago's mayor told reporters.)[41]

However, slenderness comes with a risk. Since super slenders are narrower than office buildings, they are less stable. In addition, residents, unlike office workers who are focused on work, are more sensitive to a building's sway. Some supertall tenants complain that their apartment buildings sway so much, they creak like the galleys of ships. One resident was stuck in an elevator because high wind conditions were moving and swaying the elevator cables too much. Others have complained about the ghostly swooshes and whistles of rushing air in elevator shafts. They find their lives in their cloud-like abodes disrupted by banging noises, such as a trash chute "that sounds like a bomb" upon tossing garbage.[42] At 432 Park Avenue, water leaks in mechanical floors led to tripling insurance costs. Not to mention that there is no longer free breakfast at the tower's private restaurant, run by a Michelin-star chef.

Our technology has not yet caught up to life in buildings as tall and thin as New York's super slenders. Developers trying to get to even thinner towers may be well served to remember this cautionary tale. During the 1930s, engineers pursued slenderness in bridges. The 1931 George Washington Bridge featured a 3,500-foot-long deck only 12 feet deep. Its slenderness ratio of 1:292 became a goal for other bridges.[43] In 1950, the Tacoma Narrows Bridge broke this record, with a roadway 2,800 long and only 8 feet deep, at a 1:350 slenderness ratio. However, at this extreme slenderness, the bridge's stiffness was compromised. Within a few months after opening, it collapsed in the wind.

IN 1929 NEW YORK, with demand for skyscrapers at a peak, the Metropolitan Life Insurance Building was announced. It was going to be the tallest building in world, up to 100 stories. That same week the stock market crash of 1929 occured, signaling the beginning of the Great Depression. The tower was eventually built to only a third, leaving a stump. Today, it has become a monument to economic decline.

Projects like this have led to the coining of the "Skyscraper Curse," an assumption that the tallest towers signal economic decline.[44] In tandem with the Metropolitan Life Insurance Building, the building of the Chrysler Building and the Empire State Building closely preceded the Great Depression. The Burj Khalifa was completed in 2009 just as the Dubai property market collapsed. Throughout history, the hubris of tall structures, like the Tower of Babel or the Easter Island *moai*, have seemed to tempt the gods, leading to societal breakdown.

In 2020, with a wave of new supertalls in New York still under construction, COVID-19 hit. The pandemic brought the city to an abrupt halt, including its feverish construction activity. Was the curse striking again, with New York's record-breaking skyscrapers the latest symptoms of excess, and looming economic decline?

Economists have since debunked the supposed "Skyscraper Curse." The tallest towers are not a leading indicator of economic peaks. At best, they are a lagging indicator, since construction projects of this scale have such a long lead time. And since they're so visible, they are an appealing indicator. A full skyscraper may mean economic prosperity, and an empty one may mean economic transition.

Skyscrapers are the story of urbanization, of people wanting to be in the same place at the same time. With the onset of COVID-19, those desires were questioned. The "culture of congestion," once revered by architect Rem Koolhaas, became a culture of contagion. In the age of a pandemic, tall buildings can spread infection exponentially, with people forced into social proximity in confined places like elevators. Since large buildings have far too long been configured as hermetically sealed boxes, optimized for air-conditioning and not cross ventilation, they hold even higher risk of contagion.

Density skeptics pointed to the benefits of sprawl. With people fearing public transit and social proximities, many urbanites fled to the suburbs and opted for a car-centric lifestyle. Even drive-in movie theaters had a comeback. Could this mean the end of the New York City skyscraper? "NYC must develop an immediate plan to reduce density,"[45] tweeted New York governor Andrew Cuomo, on the cusp of the crisis.

This is tragic, since cities like New York are also our greatest achievements, the root of civilization. Lewis Mumford, the noted writer on cities, described cities as "containers" that help speed up civilization. "As with a gas, the very pressure of the molecules within that limited space produced more social collisions and interactions within a generation than would have occurred in many centuries if still isolated in their native habitats, without boundaries," he wrote.[46] From their very beginnings, cities accelerated innovation and social experiments.

Ideas prosper in city centers. They condense people and offer them spontaneous and possibly fruitful encounters with others. A thousand years ago, after new plow mechanisms and crop rotations managed to sustain European towns, arrived the art of the Renaissance followed by the technological advancement of the Enlightenment.

Today, we still pick the fruits of cities. Urban areas are known to exceed rural ones in innovation and economic output. US metropolitan areas alone generate 75 percent of the national GDP.[47] In the United States, large cities produce twice as many patents per person as small cities—plus those patents have more impact.[48] Cities capture the economic and knowledge benefits of agglomeration and physical proximity in a super linear way. Double the city's size, and productivity and innovation more than doubles.

Similar to the metabolism of animals as they grow—the low metabolic rate of elephants, for example—large cities create massive economies of scale. The theoretical physicist Geoffrey West and his colleagues found that when cities double in size, they only require a resource increase of 85 percent, such as the number of gas stations and the amount of road surface.[49] This seems to be particularly true for dense cities, where resources are shared among more people, making them the centers of sustainability.

However, cities have also exacerbated problems from their very beginning. According to West, the bigger cities become, the bigger their problems. When a city doubles in size, there is a 15 percent per capita increase of violent crimes and traffic. The mathematics of social networks explain that the bigger the city, the more social connectivity per capita. However, while social connectivity is good for exchanging ideas and making financial transactions, it also speeds up more dubious activities and virus transmissions. When the pandemic and social distancing called into question dense urban life, skyscrapers took on a new meaning.

While compact cities may have contributed to the rapid spread of the COVID-19 pandemic, its opposite, urban sprawl, may have been a cause. Zoonotic diseases, which spring from animals to humans, are a result of our increasing wildlife-human interface. They are partially a consequence of the destruction of nature through deforestation and unbridled suburbanization. If we can limit our human footprint by creating compact cities, we may be able to better protect our forests and wildlife.

The genie may have already left the bottle, however. In a bizarre, forced experiment of worldwide quarantine, we learned how much technology has evolved. We learned how to work in the cloud, with platforms interconnecting remote workers. We learned how to shop and communicate online. As a result, we are now better able to change how we live. Location may be less relevant. Some people can choose to live far away from cities. Stacking up in tiny apartments, such as in a skyscraper, may no longer be as desirable.

Now engineers are responding with innovations. Elevator companies are inventing solutions to crowding, including contactless kick buttons, ultraviolet-ray disinfection systems, and better ventilation units. Sadly, these technologies will be unequally distributed. Tenants in spacious supertall towers may have contactless elevator buttons, controlled by apps. Meanwhile, residents of higher-density social housing towers will be waiting longer for slower and more crowded elevators and are thus subject to more risk.

Yet, for New York City, the pandemic may have a silver lining. History shows that it pays not to let a crisis go to waste. In the nineteenth century, New York's cholera epidemics bolstered support for a new aqueduct system, which helped spur the construction of Central Park and its water reservoir. Two centuries later, COVID-19 prompted restaurant owners, no longer allowed to serve food indoors, to reclaim parking and underutilized road spaces for outdoor dining. The streets of New York, once a conduit for cars, became home to pedestrians and diners enjoying the outdoors. New Yorkers liked it so much, they proposed to continue it in the future—crisis or not.

In the weeks after 9/11 in 2001, from New York to London, skyscrapers were doomed to oblivion. They were seen as targets for terrorists, sticking out of the urban fabric like a sore thumb. But then came the surprise! More skyscrapers rose.

Cities are laboratories for handling uncertainty and risk. They tend to find a way out of the trouble they create.

SEVEN

The Transit System That Supports Skyscrapers: Hong Kong

n 1956, the same year architect Frank Lloyd Wright proposed his Mile High skyscraper, the Federal-Aid Highway Act was passed in the United States, authorizing the construction of 41,000 miles of interstate highways. This vast system, measuring nearly twice the circumference of the Earth in length, was meant to save cities from congestion. However, its planners did not anticipate a major unintended consequence. The beltways going around the cities ended up becoming the "main streets" of countless sprawling suburbs. Ironically, Wright himself had promoted an exurban exodus through his concept of Broadacre City. It was meant as an antithesis of the city, where every family would live in a freestanding house on an acre of land. Wright wrote a book about the topic, *Disappearing City*, in 1932. He called for "a new standard of space measurement—the man seated in his automobile."

After World War II, Wright's Broadacre City came true. Americans grew accustomed to a lifestyle centered around the car. They drove to take Jack Kerouacian road trips on the asphalt less traveled. Instead of walking or taking a train, they drove to their jobs in suburban office parks. They drove to get burgers or coffee at drive-in restaurants. They drove to see movies in drive-in theaters. But someone had yet to propose a drive-in skyscraper.

Private cars have their shortcomings, including their inefficient use of space—a car driving under 20 miles per hour already takes up thirty times as much surface area as a pedestrian. The roads required to drop off the occupants and the spaces to park their cars would turn much of Wright's Mile High city in a virtual sea of asphalt parking lots amid the spaghetti of highway flyovers.

As bizarre as such a "car-scraper" might sound, many parts of cities would follow the logic of the automobile. It is estimated that about half of a modern American city's land is used by roads, streets, driveways, parking, signals, dealerships, and car-oriented businesses.[1] Some American cities have more than a third of their urban footprint dedicated to parking alone.[2] One study counted about 800 million parking spaces in the United States.[3] That's larger than the entire state of New Jersey.

As some countries laid miles and miles of asphalt to build highways and parking spaces, others built rails and created underground subway systems. Of all cities, Hong Kong has created the most seamless connections between skyscrapers and subway stations. As a result, they have been able to nourish the world's most impressive skyline with the largest number of skyscrapers. More importantly, the city stands as a testament that high density does not necessarily need to lead to congested roads if it is supported by an integrated mass transit system.

Mass transit, such as trains and subways, carry more people more efficiently than cars, giving buildings access to a larger pool of users. Not surprisingly, the world's tallest tower, the Burj Khalifa, sits next to a rapid transit station, Dubai's second-busiest metro station.

Just as large buildings benefit from infrastructure, the reverse is also true. As early as the thirteenth century, the Ponte Vecchio capitalized on the crowds crossing Florence's Arno River by lining the bridge's walkway with shops in one 276-foot-long bridge structure. Today, airports and train stations have become not just places to pass through to get on a train or a flight, but places to do business and to stay the night. What was once just a piece of infrastructure is becoming a destination.

In Hong Kong, transit developers exploit the real estate potential of their structures, capitalizing on the passenger trips they generate. Backed by this system, the city has been the laboratory for ever-higher-density mega-complexes. These developments are cities in and of themselves. At times, they house up to tens of thousands of people and offer them offices, stores, and homes, all under one roof.

In this world of subway-connected skyscrapers, people can spend their entire day indoors, without ever having to go outside. For weeks, they can crisscross in subways from one side of the city to the other. They can then walk straight from the station into the lobby of their office skyscraper, without a breath of fresh air. In this highly interiorized and controlled environment, not a square inch is left to chance. But as ingenious and efficient as Hong Kong's megastructures may be, they do come with strings attached.

FROM ANCIENT ROME to modern-day Houston, cities and buildings have been shaped by mobility devices, including our feet. Entire ancient empires, such as the Greeks, expanded around seas following the logic of warships, simply because Greek galleys were the best way to move troops and goods around. And the benefits of access to the waterways still shape cities. Today, fourteen of the world's seventeen largest cities lie in coastal areas.[4]

Transportation systems have even warped our representations of the world. The map projection used widely today is the Mercator projection, based on the 1569 Gerardus Mercator map, to facilitate maritime navigation. It unrolls the earth's sphere into a cylinder, remapping constant-bearing sailing courses on the sphere to straight lines. This causes deformation toward the outer ends, making areas close to the poles appear larger. Greenland seems the same size as Africa, when in reality it is more than fourteen times smaller.

Up until the age of the car, transportation infrastructure shaped cities for the better. Towns formed around advantageous intersections of transportation networks, such as the crossing paths of horse-drawn vehicles or walking trails. This led to the maze of meandering paths of old cities like downtown Paris, London, and Marrakesh.

Modernist Swiss architect Le Corbusier later mocked these cities. He claimed they were a result of the "pack donkey way," shaped by the whims of an animal. "The pack donkey meanders along, meditates a little in his scatter-brained and distracted fashion, he zigzags in order to avoid the larger stones, or to ease the climb, or to gain a little shade; he takes the line of least resistance."[5]

Nevertheless, these urban layouts had benefits that lasted. They are still enjoyed for the small and pedestrian-friendly scale that is conducive to walking. In addition, the undulating roads of these old urban centers build up a sense of anticipation, with pedestrians not seeing what may come around the corner.

Several of today's most romantic cities were shaped from the most efficient way of transporting goods, via water, centuries ago. Their waterways are among the most iconic places, such as the Grand Canal of Venice, with Italian gondoliers serenading "*O sole mio"*—making for a picture postcard unlikely to appear from a modern highway.

Even buildings followed the logic of urban freight. In sixteenth-century Amsterdam, merchants equipped the gables of their canal houses with hooks and pulleys. These were once used to haul goods from boats via rope into the window openings. They still do this—but instead of East Indian spices, they may haul someone's piano.

Transportation also constrained the size of urban settlements. In the fifteenth century, before the age of mechanized transport, Western Europe's largest towns, London and Paris, were each populated with about 50,000 people—similar to the earliest Mesopotamian cities from four and a half thousand years earlier, such as Uruk. These small populations did not warrant the efforts to create a tall building. Only the spires of religious structures would grace the city skylines.

Three hundred years later, backed by the mechanical power of the Industrial Revolution, city populations exploded. In 1804, after years of experimentation, a British engineer built the first operating steam locomotive, a "Puffing Devil." The regional travel railways, with the first opening between London and Greenwich in 1836, helped people commute. Cities like London were now able to radiate outward. Since steam train travel was expensive, this initially led to wealthy little villages around railroad stations, like Scarsdale not too far from New York and Brookline near Boston.

London ran into a challenge building train lines through its organic street patterns or by way of viaducts, especially through much of the world's most valuable real estate. In 1863, the city engineered a solution promoted by Charles Pearson. It created the first subway line, between Paddington and Farringdon.

The regional train and subway networks helped London expand

from a collection of villages and towns to a metropolitan area today housing as many as 14 million people. Trains increased the concentration of business and people downtown, bringing about 1 million commuters to central London every day.

Back then, the trains' coal-powered steam engines also brought terrible smog. Underground trains had "blow holes" expelling the dirty air from the underground tunnels to the streets above. These left the passengers covered in black soot.

Electricity transformed cities by bringing in electric light with around-the-clock productivity. But for cities, electric engines had an equally important impact. Trains could now go underground without "blow holes," which improved the air considerably. However, it took a while for people to believe that the electric trains were as powerful as steam-powered locomotives. In 1879, Werner von Siemens, the inventor of the dynamo-electric principle, displayed the first electric passenger train line at the Berlin Industrial Exposition. The locomotive pulled a train of only three tiny carriages, carrying six passengers each, around a 1,000-foot track. Still, a thousand people were taking a ride a day, paying twenty pfennigs each, with inquiries coming from around the world to build electric train networks. "Our electric railway is quite a spectacle here," wrote Werner in a letter to his brother.

In 1887, Frank Sprague, who would later develop electric elevators, installed the first large-scale electric railway system, in Richmond, Virginia. With his electric cars going up hills that had a 10 percent slope, he was able to prove that electric trains held up.

Backed by the power of electric mobility, stronger than horses and cleaner than steam, cities were able to grow larger and more sustainably. By the late nineteenth century in Britain, the railway employed almost five percent of the nation's workers.[6] By the turn of the century, Siemens electric trains traveled up to 128 miles per hour. Artists, photographers, and product manufacturers helped the train enter the popular imagination. They created children's toy locomotives and books made especially to be read on trains and sold in book stalls in train stations.

Skyscrapers could now more easily coexist with subways. In New York, without the blow holes, buildings could be built above subway lines. New York's Park Avenue, a once-undesirable location because of noise and pollution from steam locomotives, became prime real estate around Grand Central Terminal. This "Terminal City" included the Waldorf Astoria and the Chrysler Building. Now commuters could get to their urban destinations quicker.

Commuter-friendly streetcar suburbs mushroomed as well. Several transit agencies built suburbs in addition to railways. In London, the Metropolitan Railway bought tracts of farmland around its own tracks to develop housing. After World War I, these suburban areas were marketed as an idyllic land of cottages for Londoners sick of urban life. The developers named it "Metro-land." It later became known, as a songwriter wrote, as "a land where the wild flowers grow."

Unfortunately, progress for transportation did not always lead to urban progress. In 1908, Henry Ford introduced the mass-produced Model T, which made car travel—up until then a luxury—affordable for many. Only four years later, in New York, more cars than horses occupied the road. Soon, car washes, parking garages, and gas stations graced our cities.

The street was once a place of social gathering. It became a "traffic machine . . . a sort of factory for producing speed traffic."[7] So planned Le Corbusier in his 1929 *The City of To-morrow and Its Planning*. He advocated for cities shaped by a gridded network of highways. In a way, this made us worse than donkeys since we just followed the path of the cars' least resistance.

Ironically, streets lost their urban vitality thanks to "standards" sought out in the name of safety. Street width and parking standards gradually demanded more space, giving the largest areas of our streets to relatively few people in cars.[8] Urban sidewalks not only became unpleasant, they became increasingly nonexistent. Even when sidewalks were present, the lack of curb cuts made them hard to navigate for people with disabilities and parents pushing strollers.

As cars congested cities, they further fueled the suburban exodus. In the United States by the late 1960s, more people lived in suburbs than in cities. Federal highways and government mortgage insurance programs supported their way of life. Importantly, these new neighborhoods lacked minorities, unable to buy homes through mortgage discrimination and redlining.

A vicious cycle began. The car made suburban sprawl possible, which then made the car a necessity. Even in cities, with cars in the streets, public transit agencies reduced service and lowered costs, expelling potential transit riders. Sadly, subway and train systems did not have enough time to come to full fruition in North American cities. People were left with few alternatives to entering their cars. Low-income households can lead a fragile existence when subject to a long commute and rising fuel prices.

Instead of building rail lines, governments built highways that cut through urban communities, such as the Bronx Expressway in 1953, also known as "Heartbreak Highway." But even the widest highways are no match to subway trains' capacity of transporting people. One highway lane can carry only about 2,200 cars per hour, as determined by the *Highway Capacity Manual*, whereas typical subways can carry about 36,000 people (although Hong Kong's metro system has attained a capacity of up to 86,000).[9] As a result, car commuters were condemned to traffic jams. No wonder this led to new cultural phenomena like road rage—although some of the cultural impacts were more benevolent, like garage sales, tailgating get-togethers, and an entire industry dedicated to creating tents, bbq's, and beer taps attached to cars.

This new way of life came with an environmental toll. Cars disproportionately contribute to smog and climate change, with a car trip consuming about five times more megajoules per passenger kilometer than public transit.[10] In 2016, the global transport sector was responsible for 23 percent of the global carbon emissions, and the vast majority of this came from road vehicles.[11]

Each year road traffic kills 1.3 million people, according to the World Health Organization. The car impacted public health in other sinister ways as well. A century ago, we used to walk to school or to pick up groceries. Today, trips like this are often made by motorized means. With each additional hour spent in a car, there is a 6 percent increase in the likelihood of obesity.[12] Weight gains lead to obesity-related illnesses such as strokes and diabetes.[13]

In the 1990s, political scientist Robert Putnam found that for each additional 10 minutes spent commuting to work, there was 10 percent less involvement in community affairs, such as volunteering and even voting. In addition, informal socializing activities—for example, hanging out with friends—also dropped from 85 minutes in 1965 to 57 minutes in 1995.[14] Where Americans used to play in leagues, he noted, they were increasingly "bowling alone."

Some governments took a different route. In Denmark, the city of Copenhagen announced its new "five-finger plan" for future expansion in 1947, with a railway structuring development within each finger. With commuters having the option to take a train instead of a car, the city gradually removed parking spaces in the urban center.

These spaces were transformed into open-air cafés and neighborhood parks. Social activities increased fourfold.[15]

While cities such as Copenhagen were initially outliers, as the twentieth century progressed, city planners adopted transit-oriented developments more widely as a model for urban development. In Portland, Oregon, for example, planners brought back streetcars, strengthened in their convictions by studies proving that designing communities around transit hubs promotes the use of public transit, bringing health benefits. Nine separate studies from across the world showed that people walked between eight and thirty-three minutes more a day when using public transit, since it typically requires walking to a bus stop or a train station.[16]

While Copenhagen's and Portland's approaches work for midsized cities, the needs of a larger city like Hong Kong come with unique challenges and efficient solutions. The city today is a throwback to the nineteenth-century promise of public transit in cities, though in a new and surprising way.

HONG KONG IS OFTEN SEEN as the epitome of density, the endgame of urbanization, the ultimate concrete jungle. The city has the world's most buildings taller than 100 meters, about two thousand, more than double the number in New York City.[17]

The city itself capitalizes on its density through tourist attractions. The Peak, Hong Kong Island's highest point, offers visitors an extraordinary panorama of hundreds of high-rise towers. The Symphony of Lights, the world's largest permanent light show, illuminates the city's towers with pulsating lasers every night at 8 p.m., accompanied by a soundtrack.

There's a dystopian side to all this density. In contrast to New

York, Hong Kong's towers meet the narrow sidewalks all the way to the front, without a setback. As a result, the city's densely packed neighborhoods feel claustrophobic. With its skylines jammed with tall buildings and glaring neon lighting, the city is a filmmaker's dream. *Blade Runner* took its inspiration from Hong Kong, a perfect backdrop for a futuristic society where urbanization has run amok.

Most of the city's tall residential buildings are nondescript. Developers crammed in as many residential units into each tower as possible. They squeezed bay windows onto the building's outer faces, since they are not counted as maximum square footage by the government. The stacking of thousands of cubical bay windows

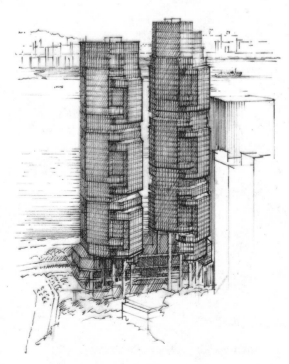

Lippo Centre, Hong Kong, Paul Rudolph, 1988

makes for puzzling facades, which seem more like infrastructure than architecture.

However, a few unique office towers stand out in a crowded skyline. The 564-foot-tall Lippo Centre, designed by Paul Rudolph, is one. With twin octagon-shaped towers in the brutalist architecture style, it is featured in the science fiction movie *Ghost in the Shell*. Architect I. M. Pei designed the 1,205-foot-tall Bank of China Tower, a structure consisting of four triangular prisms receding toward the top. It appeared in *Star Trek: Voyager* as the headquarters for Starfleet Communications. It's Hong Kong's most striking landmark—although feng shui masters believe that the building's sharp diagonal angles resemble knife blades pointing at its neigh-

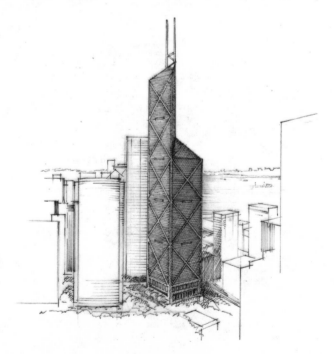

Bank of China Tower, Hong Kong, I. M. Pei, 1990

bors. This supposedly led to problems, including the financial collapse of the original owner of the Lippo Centre.

For anyone new to Hong Kong, the city's sheer density can be bewildering. I remember the many times I've felt like a sardine in a can just riding up an elevator. Or how I felt lost in the maze of crisscrossing elevated walkways and skybridges in the jumble of skyscrapers. Yet, the more time you spend in the city, the more you realize an underlying logic behind the seeming chaos.

Hong Kong operates like clockwork. You can get to almost any part of the city, with its 7 million people, within only 40 minutes. Unlike in *Blade Runner*, you don't need flying drones to do it.

You need a subway system. While an underground system by itself is not much of a novelty, Hong Kong learned a lesson from Japan. In the early twentieth century, the Japanese government nationalized 17 of the 37 existing private railway companies in order to integrate the country's growing railway infrastructure.[18] This led the private railways, no longer allowed to compete for passengers with government lines, to almost collapse.

After a typhoon there are pears to gather up. The railways were forced to diversify, most notably into real estate. A famous example is the Shinjuku railway junction, a buzzing multilevel transit interchange flourishing with department stores. Where the automobile gradually eroded private railway operators in the United States and Europe, the Japanese railway companies thrived. Many of these companies continue to do so today.

In the 1960s, Hong Kong needed a solution to its increasingly congested streets. City planners realized they couldn't just put more buses on the roads. In 1972, aware of Japan's experiences, the Hong Kong government set up the Mass Transit Railway Corporation

(MTRC) to build metro lines, in addition to developing land. Its business model "Rail plus Property" began.

The government provides the MTRC with development rights at the stations it builds. The MTRC obtains these development rights at a favorable rate, so it can recover some of the costs related to its transportation investment and make a profit.

The system has many benefits. In a high-density city like Hong Kong, there is a premium on land close to railway stations. Sometimes the values differ by several orders of magnitude. As a result, Hong Kong is one of the rare places where public transportation can actually be profitable. The transit authority capitalizes on the land appreciation that its new transit stations create. It's like Disney World, which is known as a theme park but its magical formula is really about profiting from the company's surrounding real estate.

In addition, unlike other cities, where transit corporations are separate entities from real estate developers, Hong Kong was able to more easily achieve a smooth integration between subway stops with the attached buildings. Developments quickly became larger, with a typical development having an underground subway station at the bottom, a mall in the middle, and skyscrapers on top. These rail-integrated complexes have become one of Hong Kong's basic building blocks, like the skyscraper is to Manhattan.

As a result, public transit accounts for more than 90 percent of all trips in Hong Kong, unmatched by any other city. This brings many benefits. Hong Kong has the lowest transport-related energy consumption of all developed cities. (But unfortunately, the city wastes energy in other ways, with air-conditioning accounting for 30 percent of all energy use.)[19] The city has one of the lowest numbers of people killed in car-related accidents—about 15 people per million in 2019.[20] This is more than seven times fewer than the United States.

International Commerce Centre, Hong Kong, KPF, 2010

As a result of this subway culture, Hong Kong's mental geography is not shaped by the orientation around roads, landmarks, or natural features. It is warped by the subway map. Restaurant ads prominently display the name of the closest subway stop instead of the street address.

Hong Kong's Union Square, completed in 2010, is the largest of the MTRC's building projects. It anchors more than a dozen skyscrapers in a single complex, including the city's tallest tower, the 1,588-foot-tall International Commerce Centre.

This is just the tip of the iceberg. When I first visited the project, and marveled at all the towers in the sky, an atrium window revealed a bustle below. I realized my feet were not planted on the ground; rather, I was hovering atop more buildings. I was standing on the roof of a mall, which itself was standing above one of the city's largest stations. This monolith departed from a traditional city with a neighborhood consisting of several blocks and separate buildings. It had materialized all these functions into one vast interconnected complex, complete with offices, apartments, stores, and at the time the world's highest hotel.

The project stands on Kowloon Station, at the intersection of subways, bus lines, and the airport railway. This offers a built-in convenience for residents, shoppers, and office workers. You can even check your luggage there, before you travel to the air terminal, and then go to your meeting or favorite store. Home to a total of 70,000 residents and office workers, the complex is the closest we have to a vertical city, with a population density about twenty times as dense as New York City.[21]

Hong Kongers seem to have a higher cultural tolerance for density, although this may be not as a result of choice but of necessity. There's a branch of science dedicated to the effects of density on human behavior, called "proxemics." Edward Hall, who coined the term in 1963, defined the smallest range of space—the "intimate distance"—as less than 18 inches, dedicated only to someone's closest friends, since at that distance you can sense another person's odor. He found that eye contact is avoided when this personal space is violated, in order to achieve dehumanization. Hence, one American city planner was astounded, upon his visit to Hong Kong in the 1960s, that its density did not cause pathological behavior. "Most orthodox planners," he wrote, "would expect densities of the magnitude found in Hong Kong to precipitate serious health and social problems with death, disease and social disintegration."[22]

Hong Kong defies the conventions of proxemics. In 1994, organizers of a Canto-pop concert found a special solution to the noise complaints of neighbors to the open-air stadium. They forced thousands of fans to use gloves to clap, or forget clapping altogether. The 17,500 pairs of white gloves they provided were especially commended for their strong visual impact.[23]

The city found more infrastructural solutions to alleviate density as well. It created an elevated network of footbridges, providing people with another opportunity to cross an entire neighborhood without ever touching the ground. The city's three-dimensional meshwork of footway bridges even has a life of its own. On Sundays, three hundred thousand migrant domestic workers have the day off. They use the elevated spaces for picnics, in plain sight of passing white-collar workers. These elevated footbridges could help create new class encounters—not unlike the poet Charles Baudelaire's observation of Paris's nineteenth-century boulevards.

But there is a big difference between a stroll in Hong Kong and on a Parisian boulevard. At the core of many of the MTRC's rail-integrated developments lies the mall. These retail spaces take the idea of shopping districts to the next level. Nineteenth-century arcades and twentieth-century malls usually required a trip to get there, if you didn't live in the neighborhood. Now the mall *is* the neighborhood—and conveniently not subject to dreary weather or climate conditions.

Hong Kong makes shopping seamless. You can exit the metro directly into a mall, without going into the street. You can even go from one mall to the next through interconnected footbridges. The malls are deliberately placed between all entry points into the structure, the various towers, and the subway stations. They're impossible to miss. As a result, for millions of people, entering a mall is no longer optional. Instead, it is a near inevitability. Now, people's everyday lives increasingly take them past retail corridors and shopping atria.

Besides, Hong Kong's apartments are small, its summer climate is hot and humid, so why not meet at the mall where space is plentiful and air-conditioning is free? While there, although you might not need to shop for anything specific, you might as well have a look around, and spend some money.

Situations like these achieve what shopping designers call the "Gruen transfer." This term refers to the manipulation behind the mall's undulating corridors. They aim to lure consumers from their original intentions, encouraging them to shop for shopping's sake, rather than looking for something specific. The Gruen transfer was named after the architect Victor Gruen, who designed the first prototypical mall, Southdale Center in Minnesota, in 1956. This building pioneered combining all the elements of a mall that we now take for granted—including anchor stores, escalators, and an interior atrium—all under one air-conditioned roof.

Ironically, Gruen, an Austrian immigrant, had more idealistic purposes. He envisioned the mall as a new town center to be a dense antidote to the lackluster American suburbs. The shopping component was just a piece of his vision, which also included apartments, offices, a park, and even schools. Sadly, his invention backfired when malls remained insular. They probably even nourished the frantic consumerism he was trying to reduce. Like Dr. Frankenstein, while his intentions were for the greater good, he ended up creating a monster.

"Those bastard developments," Gruen exclaimed during a revisit of his American creations. They had discarded Gruen's intended community functions and refused to surround themselves with other buildings. Instead, they were enveloped with "the ugliness and discomfort of the land-wasting seas of parking."[24]

Unlike their suburban counterparts across the Pacific Ocean, Hong Kong's urban malls lie closer to Gruen's original vision. They are part of a larger community, surrounded by buildings with different uses. Even better than Gruen's imagination, they are integrated into subway systems.

They are also extremely profitable. Hong Kong has the world's highest-priced retail space. In 2019, its top retail area rented for an average yearly $2,745 per square foot, $500 more than New York's Fifth Avenue.[25] That's half a million dollars a year for an area the size of a parking space!

Meanwhile, in the United States, there are many sightings of "dead malls." There, observers herald the end of the mall, with "ghost boxes" (empty anchor stores) and "label scars" (the monster that refuses to die: still-existing signage of a previous tenant).

Conversely, in Hong Kong, Gruen's creation has reached an entirely new order of magnitude. The city even features "vertical malls" like Hysan Place, stacking shopping floors 17 stories, as high as Big Ben. In Langham Place, architect Jon Jerde innovated with "expresscalators"—high-speed escalators able to shoot up four stories in a single shot within only 55 seconds. These are "irrigating" the difficult-to-reach higher floors with an incessant stream of shoppers. They whisk shoppers through spectacular atria, like the soaring, nine-story glass atrium of Langham Place. The 19-story mall of MegaBox even has multistory atria in the shape of ovals. These give visitors a mental "restart": a type of psychological break from the concrete jungle's average ceiling height.

All of this makes Hong Kong a city of malls. With about one mall per square mile, it has the world's highest concentration of malls.[26] With China expected to become the world's largest consumer economy, the fate of Gruen's invention will take another turn. Developers in mainland China and other places now learn closely from Hong Kong's megaprojects.

Hong Kong's megaprojects may qualify as true vertical cities. But many of the malls' exterior edges have blank walls, with all the charm of a prison, instead of shop fronts that lend excitement to the street. Meanwhile, the "public spaces" of atria and interior corridors feel manufactured. There is none of the messy vibrancy of Paris or London, despite the occasional exhibitions and ice rinks. The outdoor spaces are typically privately owned public spaces (POPS), which in reality are not so public at all. Unfortunately, developers like to hide the open spaces on rooftops, making them accessible only by passing through the mall. In short, the trick is to "create the POP to get people to shop."

At the POPS in front of the mall, extensive use by the public is discouraged. Hong Kong's Times Square, another vast podium-tower project, has deliberately provided uncomfortable chairs for sitting. "Overzealous" guards further prevent people from lingering too long. Placards detail the outlawing of demonstrations, musical instruments, and pets. The point being that there is space for the sole purpose of saying it is there.

These and other issues raise difficult questions about the mega-corporation behind the megastructure, such as whether or not real urbanism can exist under the rule of property managers.

At worst, Hong Kong's megaprojects represent a dystopian future of a city hit by the "shop apocalypse."

Still, Hong Kong's mall-skyscraper hybrids make for highly convenient living, thanks to their integration in mass transit. These buildings are examples of a holistic design methodology, a fusion of city planning, public transportation, and skyscraper architecture.

So, three years after my initial dismay about Union Square, with this more nuanced view, I ended up staying in one of the project's apartments. Within a few days, I came to begrudgingly appreciate the thing.

Ironically, when I returned to the city a few years later to launch a book about Hong Kong's urban development, *Mall City*, I was interviewed by a reporter, a Hong Kong native. She wanted to interview me in the mall in one of the megaprojects.

The mall was a disorienting maze. We both ended up getting lost.

"LOCATION, LOCATION, LOCATION" goes the well-known broker's mantra. It refers to the phenomenon in which two buildings may be identical. However, one may be worth a lot more than the other simply because it is in a different location. While location matters, it is not precise enough. The mantra should be "access, access, access."

A place derives value from its access to schools, restaurants, grocery stores. Access improves as neighborhoods blend more and more uses or as they densify, for instance with tall, mixed-use buildings. Physical proximity has long been of the essence for access. However, better mobility systems can also improve access. With autonomous vehicles, hyperloops, and aerial ridesharing in the near future, we may be up for a wild ride! Plus, along with it, land that was previously less desirable will become the development hot spots of tomorrow. New mobility systems will change our cities and buildings once again.

In 1994, Italian physicist Cesare Marchetti described a principle, now known as "Marchetti's constant." He found that, in general, people are willing to commute for about one hour a day, or a half-hour one way. Since Neolithic times, people kept the average time spent for travel the same, even when transportation advances allowed them to increase their distance. This idea has had far-reaching effects on our cities.

Before the Industrial Revolution, our urban settlements were constrained by a 30-minute walk. This explains why most premodern cities grew to a maximum radius of about one mile. The historic city of London is nicknamed Square Mile for this reason. Even mighty ancient Rome was limited by people's feet, with an urban footprint of about two miles in diameter. The first forms of mass transit didn't change this radius much, like the horse-drawn carriage lines in Paris in 1662, since hooves did not move people faster than feet.

With electric streetcars in the late nineteenth century, commuters could travel about 4 miles in half an hour. They often chose to do so. They could now live on cheaper land that was previously worthless because it was inaccessible, which helped reduce the price of housing. In the late twentieth century, the distance of a commute expanded even more with the faster speed of expressways. Metropolitan areas grew up to a vast 40-mile radius.

Today, a prime "location" may be even more of a moving target. Various companies are developing hyperloops, sealed tubes with magnetically levitated pods that could convey people in a highly energy-efficient way. Capsules carrying people can reach a top speed of 760 miles per hour, around the sonic barrier. This may lead to the explosive growth of metropolitan areas. You could commute from your office in downtown Los Angeles to Las Vegas, where prices are lower, and live in a bigger home. The race will be on for the first hyperloop-integrated skyscraper.

Autonomous vehicles (AVs) could desensitize people to distance, possibly lengthening Marchetti's "constant" of commuting time. Without having to attend to the wheel, you can now sleep in your car, or conduct online meetings there. Perhaps you can live a little farther away and buy a cheaper home? In other words, cities will increasingly sprawl out. The increase of telecommuting, accelerated by COVID-19, may have a similar impact. Workers may choose to live farther away when they require fewer commutes to their downtown office.

Self-driving cars may have more positive implications on downtown cores. They can liberate a lot—a *parking* lot. When the car drops you off, it likely won't be parking. It will pick up someone else, since many of us may be sharing cars. The obsolete city parking lot can then be repurposed. In New York, some predict this AV-reduced parking demand can free up roughly 900 city blocks, the size of six Central Parks.[27] This means a lot of potential for parks, or for skyscrapers.

Already, parking garages are being "future proofed." Some developers are departing from building typical garages with low ceilings and ramped floors. Instead, their garages have ceilings with demountable ramps that can be converted to other uses later on, such as office space. It may cost more to build this way. Yet, if parking demand reduces in the future, the building can have a new life. Ironically, this is a return to our first parking garages, like Garage Rue de Ponthieu, built in Paris in 1905, which had high ceilings and a skylight.

There may be even more futuristic types of changes that are hard to imagine today but might be possible over the next decades. Autonomous and electric vehicles are likely safer for pedestrians, because of their lack of exhaust emissions and human unpredictability. This may allow a more peaceful coexistence between people and vehicles in streets. In some ways, the future of the automated street could be a throwback to the era before the car. In those days, the street buzzed with social activity, where people would shop and meet. Perhaps a street for safer, autonomous vehicles has the potential to return to its more sociable days.

With safer conditions for pedestrians, streets can remove their curbs. The autonomous streets could look a lot like the "shared streets" that already exist in Europe. For decades, entire Dutch neighborhoods have removed curbs to increase the perception of risk for automobilists, leading to slower traffic. Studies prove this has increased social interaction and childhood play,[28] and today two million Dutch people benefit.

Streets may also become more flexible. With AVs, the number and direction of lanes during the day can change, instead of being fixed, because software will determine where and how cars will drive. This allows for a more flexible use of roadways. Outside of peak traffic hours, lanes could close to give way to pedestrian space.

The classic highway interchange may get a makeover as well. NASA considers the unused areas around highway cloverleafs ideal landing spots for vertical take-off and landing (VTOL) aircraft, where sound is drowned out by speeding cars. These will become desirable sites when aerial ridesharing serves new routes between frequently commuted routes with heavy car traffic. Because, inside cities, people have yet to accept their urban skylines flocked with the noise of buzzing drones.

It is easy to be seduced by the speed of hyperloops or the aerial sophistication of a flying drone. While hyperloops are faster, they probably won't come close to the transport efficiency of the good old subway. Hyperloops can carry 3,360 passengers an hour.[29] A single subway can carry more than ten times as many people. (While it might not carry as many people, a hyperloop's higher speeds would prevent people from otherwise taking a more carbon-intensive ride.)

If Marchetti's constant continues to hold, building new developments around faster forms of mobility may further encroach on untouched land, destroying habitats and threatening ecosystems. But there is another option.

Access is a function of proximity and mobility. Instead of focusing on new mobility, we can increase access with proximity, by more densely packing activities or with a larger variety of uses. Hong Kong's subway-connected megastructures give people the most access to the most destinations.

Meanwhile, developers are making America's old-school suburban malls more like their dense counterparts across the Pacific in Hong Kong. In 1950, the Northgate Mall in Seattle opened as one of the first suburban mall–type shopping centers. Today, builders are tearing up the asphalt parking lots and erecting residential and office towers, while a new light-rail stop will connect the development to the city. The once archetypical suburban mall is increasingly adopting urban characteristics.

At the same time, in Hong Kong, the MTRC's engineers continue to improve their system. They are building new underground lines, running more train trips per line with more cabins, and drilling stations deeper into the ground. As they are increasing the system's capacity, they are improving service by adding breastfeeding facilities, baby-care rooms, cleaning robots, and robots providing directions 24/7.

This explains why the marker of a tall building in Hong Kong is not a fancy address, prominent landmark, or natural scene, but an exit into the underground, thus revealing the love affair between the skyscraper and the subway station. Born from the same engine (the electric motor), these twin inventions relate to each other like yin and yang, like lightness and darkness—they are the force of more sustainable urbanization.

EIGHT

The Greening of Vertical Cities: Singapore

ommon knowledge holds that with urbanization comes the concrete jungle, a crowded forest of skyscrapers shrouded in air pollution while surrounded by filthy rivers. The nineteenth-century epitome, London, had its River Thames filled with putrefying carcasses, human waste, and rotting sludge. One hot summer in 1858 exacerbated the already foul-smelling scent to such an extent that it went into the history books as the Great Stink. Today, science fiction often depicts the "urban" as a dystopian, dense city of asphalt, where not a single tree or blade of glass is to be found. Think *Blade Runner*, its urban scenes permanently dark and over-run by buildings.

The underlying assumption is that with economic growth comes environmental decline. And this is not too far from the truth. Even a nightly satellite view of the Earth shows urbanized areas as bright swaths of light. From above, darkness is a good thing, referring to nature, unlit. Inside these supernovas lie only a few patches of pitch blackness. The bigger the city, these images show, the less green there seems to be.

Some cities are fighting back on this narrative. Singapore stands as an example on how to avoid this "progress trap." As its economy has grown, so has its green cover. Today, trees and vegetation cover over 50 percent of Singapore's landmass, more than any other major city. Within all of this green acreage, the city still manages to squeeze in more than six thousand high-rises, giving it by some measures one of the world's most impressive skylines.[1] And even the city's skyscrapers feature verdant walls and lush roof gardens.

Singapore has also pioneered the use of vertical farms, bringing skyscraper principles to greenhouses, with vegetables and lettuce stacked on rotating Ferris wheels. Even those who are just transferring through Singapore's airport can experience the city's greenery: Terminal 4 features an indoor jungle with the world's tallest indoor waterfall. Singapore did not start out this way. Back in the 1960s, the city was polluted and lacked proper sanitation. Instead of accepting this as the cost of development, planners used it to rethink progress in inventive and green ways.

All of this shows that cities and skyscrapers don't have to alienate people from nature. Soon, Singapore may become a forest of lushly vegetated skyscrapers, giving a new spin to the term "concrete jungle."

AS EARLY AS THE BEGINNING of cities, those cities that debased their natural surroundings did so at their peril. In 3,000 BC, Uruk was more densely populated than modern-day New York City, with 80,000 people crammed into an area of a little over two square miles.[2] This crowded capital had to continually expand its irrigation system to feed its growing population. In Sri Lanka 2,500 years later, the city of Anuradhapura had a similar problem. It was also growing constantly, and like Uruk, it relied heavily on an elaborate irrigation system.

As Uruk grew, its farmers began chopping down trees to make space for more crops. In Anuradhapura, however, trees were sacred. Their city housed an offshoot of the Bodhi tree under which Buddha himself was said to have attained enlightenment. Religious reverence slowed farmers' axes[3] and even led the city to plant additional trees in urban parks. Initially, Uruk's expansion worked well. However, without trees to filter their water supply, Uruk's irrigation system became contaminated. Evaporating water left mineral deposits, which likely rendered the soil too salty for agriculture.[4] Conversely, Anuradhapura's irrigation system was designed to work in concert with the surrounding forest. Their city eventually grew to more than twice Uruk's population,[5] and today Anuradhapura is one of the world's oldest continuously inhabited cities and still cares for a tree planted over 2,000 years ago.

This tale of two ancient cities still rings true today. We may think nature is unconnected to our cities. But trees have always been a crucial part of flourishing urban spaces. At their core lies an impressive biotechnology. A mature, healthy tree can have a few hundred thousand leaves, each an instrument of photosynthesis. These porous leaves purify the air by trapping carbon and other pollutants, making them essential in the fight against climate change. In addition, trees act like a natural sponge, absorbing stormwater runoff before releasing it back into the atmosphere. The webs of their roots protect against mudslides while allowing soil to retain water and filter out toxins. Roots help prevent floods while reducing the need for storm drains and water treatment plants. "In some Native languages," the ecologist Robin Wall Kimmerer has written, "the term for plants translates to 'those who take care of us.'"[6]

Trees even support each other collectively through mycorrhizae, fungal bridges that move carbohydrates between trees. This is "Earth's natural internet," Kimmerer has noted. "A kind of Robin Hood, they take from the rich and give to the poor so that all the trees arrive at the same carbon surplus at the same time ... the trees all act as one because the fungi have connected them."

Humanity has been uncovering these arboreal benefits for centuries, and for different purposes. By 8,000 BC, our ancestors, before they built permanent settlements, likely planted and cultivated fruit trees to help them gather food.[7] When our predecessors established our first cities, they re-created nature for the purpose of recreation and entertainment. Around 600 BC, the Babylonians were said to have built the mythical Hanging Gardens, a structure with ascending tiered gardens, classified as one of the Wonders of the World. Although there has been no definitive archaeological evidence, a neo-Babylonian king supposedly built it for his Median queen, who missed the rolling hills of her birthplace.

Our ancestors also planted trees for religious purposes. Each of ancient Rome's hills featured sacred groves, *horti*. Cutting them down would come with severe penalties, including the ultimate price, death. In the thirteenth century, Kublai Khan required trees for practical reasons. He specified them along both sides of public roads into Beijing, spaced two paces apart. The trees provided shade in summer and acted like road markers in winter, when the road was invisible due to snow.[8]

Trees played an increasingly important role throughout the development of modern Europe. During the Renaissance, the military used trees to fortify city walls, such as the poplar trees on Lucca's walls in Tuscany. The roots helped stabilize the earth within the brick structures. With a scarcity of trees within city walls for common people, these vegetated bulwarks became a popular place to walk.[9] The Lucca trees still remain in high demand today.

These trees provided quite the spectacle. Upon his view of the trees on the city walls of Antwerp in 1641, a few rows of lindens, the English author and gardener John Evelyn wrote: "There was nothing about this City, which more ravished me than those delicious shades and walks of stately Trees, which render the incomparably fortified Works of the Town one of the sweetest places in Europe." Unfortunately, armed conflicts made many trees casualties of war. They often ended up as stakes in palisades.[10]

During the Baroque period of the seventeenth century, the elite once again aspired to create urban green spaces for recreation. City officials made tree-covered city walls accessible to people as a shaded backdrop for socialization and mingling. Trees were planted inside cities and in spectacular gardens, most famously in Versailles. Here, rows of pines, elms, and fruit trees lined avenues called *allées*. They stood beside exuberant beds of flowers—some of which were in so much demand that this led to the Dutch tulip mania in 1637, the world's first speculative bubble.

During this time, Amsterdam became known as a city with abundant tree-lined canals. This sight astonished visitors; the British diarist John Evelyn described Amsterdam as "a City in a Wood": "Nothing can be more pleasing, especially being so frequently planted and shaded with the beautiful lime-trees, set in rows before every man's house."

While the reasons behind these urban trees remain uncertain, the saplings were most likely planted for both aesthetic and practical purposes. The Dutch during this time experienced a "golden century" with a new wealthy and cosmopolitan class of merchants, who may have aspired to trees as a symbol of prosperity. The trees also aided in the stabilization of the quayside. The trees they planted, poplars and lindens, were hydrophytic trees, which helped suck out water from swampy marshes and wetlands through solar energy.

The Dutch cherished their valuable trees, protecting their roots with wooden boxes. Stringent regulations punished the damaging of a tree—according to one 1454 law, this came at a hefty monetary sum, or your right hand.[11]

Where the Dutch placed trees along canals, in Paris they were placed along special streets. Queen Marie de Medici, after the assassination of her husband, French king Henry IV, built the first allée of city trees outside of a city wall. She placed them along the banks of the Seine. These tree-lined boulevards even led to a new word in the French dictionary: *promenade*, a walk with the purpose to "see" and to "be seen." These streets for strolling would provide the etymological root to the introductory parading of guests at a formal ball, and the American high school "prom."

Urban trees became more commonplace when French global imports grew wealth and a new class, the nouveau riche, eager to assert their heightened standing. They took up aristocratic sports like lawn croquet and lawn bowling. But, where French royalty played in secluded gardens, the nouveau riche took to the *allées* in cities. As a result of the newfound popularity of these games, tree-lined streets flourished, albeit unintentionally. They were a status symbol projecting success, wealth, and lawn croquet. Sadly, today there is still an unequal distribution of trees for people in areas with a lower socioeconomic status.

In France, the tree benefited as a symbol of fertility in folklore as well. After the French Revolution separated the aristocracy from their power (and sometimes their bodies from their heads), a "Commission of Artists," as part of the rebellion, aspired to a different Paris. They wanted to replace the symbols of the monarchy and the church. They adopted the "liberty tree," and lavishly planted them in front of major buildings, such as former palaces, monasteries, and prisons.[12]

It wasn't until the nineteenth century that city planners embraced trees more systematically in urban areas. Between 1853 and 1870, Napoleon III commissioned Baron Haussmann to undertake a massive citywide renovation. Haussmann diligently created new parks and a network of wide boulevards, such as the famed Champs-Elysées, all lined with equally spaced trees. The saplings were an essential part of a host of upgrades, including sidewalks and streetlights. In addition to being associated with the wealth of royal gardens, they were perceived to have health benefits, purifying the air and preventing potential disease. Their visual impact also served the interest of the young republic. The sight helped to strengthen the linearity and perspective of the long boulevards, which focused on key monuments, like the Arc de Triomphe.

An army of pruners climbing up with spikes pruned large elms every year to keep them perpetually young. With fascination, the *Scientific American Supplement* reported in 1898: "This work has to be performed by pruners of recognized skill.... A good pruner should be nimble and prudent, not be subject to vertigo."[13]

In contrast to the aristocratic allées of Louis XIV, the boulevards became tree-lined promenades for everyone. As French impressionists like Claude Monet depicted urban scenes with trees, other cities also wanted their own green boulevards. Emperor Franz Joseph I of Austria sought to modernize Vienna. This meant taking down the old and obsolete medieval city wall. Inspired by Haussmann, he replaced the wall with a green ring around the old city, featuring a wide boulevard with parks and trees. The Ringstrasse became Vienna's most famous street and Sigmund Freud's favorite place to stroll.

In the eighteenth century, entrepreneurs realized that trees and parks could increase property values. The green quad was born. It rose to prominence most of all in Great Britain, where it was attributed in part to the influence of the grassy Cambridge and Oxford campuses. Developers built homes and businesses around rectilinear urban parkland. With the dominance of the British Empire, such quads started to spread. In 1733, they made their way to Georgia. Colonel James Oglethorpe planned the city of Savannah following English city planning principles. His egalitarian belief was that each homeowner should have access to open areas, which led to a city flush with green quads. He ensured that no neighborhood was more than a two-minute walk from a park. As one young woman observed in 1833: "In laying out the city every other square has been left as an open one, enclosed with a railing, laid out with walks and planted with shade trees and rustic seats arranged in them all about. These manifold grassy parks, or lungs of the city as I heard them called, are very picturesque and inviting, and highly suggestive of health and comfort."[14]

Sadly, with urbanization, forests were cleared for agriculture, and poorer inhabitants felled urban trees for fuel. Thomas Jefferson, already suspicious of urban dwellers, denounced the trees' removal as "a crime little short of murder." "How I wished that I possessed the powers of a despot!" Jefferson once exclaimed at a dinner party. "I wish I was a despot that I might save the noble beautiful trees that are daily falling sacrificed to the cupidity of their owners, or the necessity of the poor. The unnecessary felling of a tree, perhaps the growth of centuries . . . pains me to an unspeakable degree."[15]

During the rapid industrialization of the Second Industrial Revolution, exponentially growing cities became increasingly dirty. People assumed that "miasmas," the smells from polluted water and sewers, were causing disease. Trees, they thought, would sanitize the atmosphere and improve public health. In addition, parks were seen as vital tools to keep the lower classes from less savory pursuits. "Your Committee feel convinced that some Open Places reserved for the amusement (under due regulations to preserve order) of the humbler classes ... would assist to wean them from low and debasing pleasures," such as "drinking houses, dog fights, and boxing matches."[16] This led to places such as Victoria Park, the first purposely built park in London.

In New York, by the middle of the nineteenth century a large part of the European migrant population was housed in squalid living conditions in tenements. The city's elite called for a large park. Up until then, one of the few places for the public to stroll amid art, sculpture, and rolling hills was the city's cemeteries. A design contest was held for a city park. In 1858, Frederick Law Olmsted's plan was chosen, leading to Central Park. Olmsted conceived the park as a natural landscape in the British landscape tradition. Olmsted had worked as a sanitary officer during the American Civil War, and early on recognized the benefits of trees and vegetation. "It is a scientific fact that the occasional contemplation of natural scenes ... is favourable to the health and vigour of men."[17]

Other than Central Park, Manhattan had few trees by the end of the nineteenth century. Photos from around that time make New York seem like a movie set, without any vegetation. Lacking trees to provide shade, buildings absorbed solar radiation during the deadly summer heat waves. Combined with the period's poor sanitation, the oppressive heat made the city a breeding ground for bacteria like cholera. Physician Stephen Smith authored a study finding that high temperatures correlated with childhood deaths from infectious disease. "Heat develops conditions in the tenement houses exceedingly fatal to child life," he wrote.[18] He realized trees could save several thousands of lives per year, by mitigating the excessive heat during summers. In 1873, he introduced a bill to the New York state legislature to establish a Bureau of Forestry, which would cultivate street trees. Unfortunately, Smith's bill wasn't approved until decades later. By then, the city didn't have the money for the trees. Taking fate into his own hands, in 1897 Smith joined a group of private citizens called the Tree Planting Association. The group would plant trees in front of public schools and along tenement blocks.

But the outflow to suburbia had already begun. Suburbs offered a green dream, a detached suburban home on a green lot on a tree-lined residential street in a neighborhood featuring large parks. Some of these suburbs paradoxically offered green only after developers had felled forests and natural areas to make way for suburban tracts. In the United States, "white flight" to greener pastures left the poor behind in gray cities.

Suburbanization separated cities from precious tax resources, which led to fewer parks. However, some progressive city governments, like Portland, Oregon, planned citywide park systems. In 1903, John Olmsted, the stepson of Central Park's designer, was hired to design Portland's green network. He suggested turning the city's forested hills into a park. He noted: "Some people look upon such woods merely as a troublesome encumbrance standing in the way of more profitable use, but future generations will not feel so and will bless the men who were wise enough to get such woods preserved."[19]

Today, Portland's Forest Park is the world's only city wilderness park. It preserves the region's natural biodiversity, making the park home to various local plants, 112 bird species, and 62 species of mammals,[20] including deer, bobcats, and gophers. Because the city has interlinked its parks into a network, wildlife can migrate more easily, which further promotes biodiversity. This shows that urban trees don't just benefit people.

Some European cities were also wise enough to preserve parkland. After World War II, cities rapidly expanded. Copenhagen's five-finger plan included "green wedges" between each of the five arteries for new development. This increased the city's agricultural and recreational land and reduced pollution. Most cities, however, did not expand with nature in mind. Bulldozers excavated hillsides, asphalt machines paved over grass, and cement trucks filled every culvert. By the late twentieth century, with climate change increasing the intensity and frequency of storms, problems came. Cities started to flood.

Currently, little by little, green has made a comeback. Some cities sought to undo part of the damage done and restore their rivers. In Seoul, Korea, an expressway and concrete paving had covered up a stream that once flowed through the city. In 2003, the mayor decided to take down the forty-year-old elevated expressway, restore the polluted stream, and create a park along its banks. The Cheonggyecheon stream is now a thriving recreation space. Species such as fish, birds, and insects have returned, while the stream also does its part to alleviate the urban heat island effect, reducing the temperature of its adjacent areas by 6.5 degrees Fahrenheit (3.6°C).[21]

Green projects like this also occur to let nature do the work of pipes, pumps, and process plants. Cities increasingly build bioswales and rain gardens to replace stormwater pipes and permeable pavers to edge out concrete sidewalks. Together, these measures work as a "green infrastructure" network to capture stormwater and filter water. All of this can help save costs on stormwater and paving systems. For instance, replacing Seattle's streets with permeable pavement helped cut paving costs in half.[22] Seattle also has a city ordinance requiring 30 percent of land parcels in commercial zones to be green and vegetated with planting. In addition, the city is providing bonuses for rainwater harvesting, larger trees, green walls, and green roofs.[23] As these green initiatives help reduce stormwater runoff, they also save energy, since drinking water and wastewater systems, usually operated by local governments, can represent about a third of a municipality's energy consumption in the United States.[24]

Trees and vegetation are also moving to roofs, promoted by green roof ordinances. Zurich was one of the first cities to implement a law that mandates all flat roofs in the city, except for terraces, to be green surfaces. In Germany, about 14 percent of all roofs are greened.[25] The city of Hamburg even enacted a policy to top 70 percent of all suitable roofs with vegetation.

This is a throwback to the green roofs of centuries past, like the Viking houses in Newfoundland. Until the nineteenth century, Norwegian log homes had been covered in turf, at times mixed with flowers and small trees, in order to provide the roof with thermal insulation. Today, green roofs are also praised for increasing the life span of roofs, since the vegetation protects roof membranes from ultraviolet radiation and temperature fluctuation.

Green roofs are increasingly home to vegetables. Urban agriculture, whether on roofs or in community gardens, has been making its way into cities as a way to reduce "food miles": the energy wasted in shipping food from far away to your local grocery store. With fruit trees beginning to be part of cities, we have come full circle to our nomadic ancestors, whose lives centered around cultivating food, including trees known for their edible fruits.

Trees have come a long way from city walls to croquet allées, and from vegetated quads to building roofs. They are reaching full fruition in Singapore, the city that most embodies the potential of vertical green, with the most tree-covered skyscrapers.

ON MY FIRST VISIT to Singapore, it struck me how remarkably different the city looked from Hong Kong, despite its underlying similarities. Both are former British colonies with a comparably sized land area, and with similar population numbers and industries. Yet they could not be more different. In Hong Kong, closely spaced skyscrapers and underground infrastructure make it difficult for trees to grow. The city has narrow streets and crowded sidewalks, with skyscrapers blocking sunlight. With all the cables and pipes in its soil, very few trees survive in the city's urban core. This contributes to the city's dangerously poor air quality, which can cause bronchitis and diminished lung function.

In contrast to Hong Kong's frenetic concrete jungle stands Singapore, a green oasis of calm. At the root of these two different destinies lies an opposing governing approach. Postcolonial Hong Kong was largely market-led, built by developers without too much of a grand plan. Singapore is top-down, led by the strong hand of a philosopher king, where nothing was left to chance. Both cities prospered, but in entirely different ways. Hong Kong became a public transit mecca, Singapore a city with a green thumb.

These differences can be traced back to 1965, in the aftermath of British colonial rule, when the Malaysian parliament voted unanimously to expel Singapore from the Federation of Malaysia. In this watershed moment, Singapore became the first nation-state to unwillingly gain independence. This left the small country, lacking natural resources, in a tough position. The new country's prime minister, Lee Kuan Yew, had major challenges to solve. "I searched for some dramatic way to distinguish ourselves from the other Third World countries," Lee said. "We struggled to find our feet."

"To achieve First World standards in a Third World region, we set out to transform Singapore into a tropical garden city," Lee decided. "Greening raised the morale of people and gave them pride in their surroundings."[26] In 1963, before independence, Lee had launched the first Tree Planting campaign. He planted the first tree himself, a *Cratoxylum formosum*, known for its light pink, cherry blossom–like flowers. After independence, he strengthened these efforts. He launched the Garden City campaign and an annual Tree Planting Day to beautify Singapore. Lee chose the month of November, since this is when saplings need the least amount of water, at the cusp of the rainy season. In 1974, Singapore had 158,000 trees. Forty years later, it had 1.4 million.[27]

In 1973, Lee set up the Garden City Action Committee and sent out green missions across the globe. "Our botanists brought back 8,000 different varieties and got some 2,000 to grow in Singapore."[28] Lee personally picked *Vernonia elliptica*, an unusual choice, since it has no flowers and, if unkempt, looks like a weed. But the city's gardeners used the species widely to decorate the walls of unsightly buildings, bridges, and overpasses.

Lee, nicknamed "Chief Gardener," enticed the leaders of his neighboring countries to go green as well. "I encouraged them, reminding them that they had a greater variety of trees and a simi-lar favourable climate." This would lead to a green race, with neigh-boring countries trying to "out-green and out-bloom" one another. "Greening was positive competition that benefited everyone—it was good for morale, for tourism, and for investors," Lee assumed.[29]

Greening also became about survival. Singapore is a country

the size of a city. With about 6 million people, it has the same pop-ulation as Denmark, but in an area only half the size of London. As a result, the nation is dependent on neighboring countries, like Malaysia, for things as basic as water. However, Lee knew that his neighbor could cut off Singapore's lifeline, fresh water, during times of conflict. Malaysia's president once said, "We could always bring pressure to bear on them by threatening to turn off the water."[30]

To avoid relying on other countries, Singapore needed to be self-sufficient within its own compact footprint. Having to capture rain-water, it could not afford to leave its rivers polluted, as so many other countries have done. Singapore, in the name of self-sufficiency, had no choice but to go green.

In 1963, Lee consolidated different entities to set up a national water agency. For ten years, the agency toiled to clean up the rivers, which up until then were an open sewer. Public officials relocated factories and farms and built water reservoirs, planning to collect and reclaim stormwater in the city. "By 1980, we were able to provide some 63 million gallons of water per day," Lee stated, "about half of our daily water consumption then."[31]

Today, Singapore features a myriad of water reservoirs, rooftops, parks, roadways, and sidewalks to all capture water. Two-thirds of its surface is a water catchment area. An elaborate system of chan-nels, tunnels, and pumps then moves the water to treatment plants, all controlled by microprocessors.

Parallel to greening Singapore, Lee wanted to get people to own flats. Homeowners, he assumed, would have a bigger sense of belonging than tenants. The city's Housing & Development Board (HDB) would build low-cost housing that citizens were allowed to rent and then purchase with their pension funds. Today, 88 percent of all Singaporeans are homeowners, among the world's highest

home-ownership rates. It's worth noting that the system deliberately disadvantages same-sex couples and excludes several hundred
thousand migrant workers, who are living in crowded dormitories.

With limited land supply and rapid population growth, Singapore had no choice but to build up. It needed to house everyone in
skyscrapers. This transition to high-rise living did not come easily,
especially for pig farmers, Lee noted. "Some were seen coaxing their
pigs up the stairs!"[32]

The groundwork of Singapore's new green skyline was laid. As
the state mandated green policies and high-rise buildings, it was
only waiting for nature to intertwine with the skyscraper. Defying
the negative stereotypes around public high-rise housing, the city's
skyscrapers became sleek, modern, and increasingly vegetated. In
2009, the HDB completed the Pinnacle@Duxton, the world's tallest
public housing project. It features seven 50-story towers interconnected with elevated landscaped gardens, allowing residents a daily
jog among palm trees, 500 feet above the ground.

Cheong Koon Hean, who served as head of Singapore's national
urban planning authority, continued the city's green arc in the last
two decades. She infused the city's new central business district,
the Marina Bay, with an urban water reservoir and a 250-acre
botanic park, the Gardens by the Bay, featuring 18 "super trees," vertical gardens as tall as a ten-story building. Architect Moshe Safdie
designed the district's signature project, the Marina Bay Sands, an
integrated resort built of three 57-story hotel towers topped by a
connecting 1,120-foot-long SkyPark. Perhaps the most innovative
is how all this greenery coexists in an urban center, along with skyscrapers. "We intersperse parks, rivers, and ponds amid our high-
rises," Cheong said.[33]

The city passed building regulations with an important impli-

Pinnacle@Duxton, Singapore, Arc Studio Architecture + Urbanism, 2009

cation for tall buildings. If developers build on an open space, they must replace it with green elsewhere in the project. Through LUSH incentives, or "Landscaping for Urban Spaces and High-Rises," developments can create sky terraces and gardens to satisfy these requirements. The authority even encourages developers to include plants with a higher leaf area index, accounting for how some species have more leaves than others, and hence more benefits.[34] All of this helps spawn even more green cover. In the Marina Bay, for instance, developers needed to replace 100 percent of the landscape lost on the ground due to their buildings with greenery in the sky.

With all these LUSH requirements, the city has become a breeding ground of truly green buildings. Just south of Marina Bay lies Marina One, a development with several towers serving 20,000 residents and office workers. At its core sits a terraced, multilevel garden with snaking wooden walkways, home to more than 350 species. Unlike typical buildings, floors have deep planting beds for drainage, absorbing water during times of tropical downpours.

Marina Bay Sands, Singapore, Safdie Architects, 2010

Just west of Marina Bay, the Parkroyal Collection Pickering hotel envelops hotel guests with trees and vegetation. Every four levels, tropical plants are draped from sky gardens, featuring palm trees and blooming frangipanis. Another skyscraper, the Oasis Hotel Downtown, is encaged by a red aluminum mesh, which 21 species of creepers will gradually fill in. With each type of plant better able to survive depending on the mesh's solar orientation and shade, the creepers and flowers will make for a unique pattern yet to come. The mesh covering almost the entire skyscraper will replace a record of more than 10 times the green area lost on the ground. Meanwhile, the city's record holder for the largest vertical garden is the Tree House, a 24-story condo tower in the West Region of Singapore. The green wall covers one of the building's sides entirely, measuring almost 25,000 square feet, about the size of five tennis courts.

Parkroyal Collection Pickering, Singapore, WOHA, 2013

Singapore plans to use all this greenery to offset its fundamental fault. The city came at the cost of its tropical forest. Only 0.5 percent of the nation's primary forests remain. Urbanization impacted the climate, with urban areas up to nine degrees Fahrenheit warmer than rural areas.[35] The city's newly planted trees and green walls will help cool buildings, provide shade, and reduce outdoor temperatures. Hopefully, this will encourage people to walk or take a bus, instead of taking a climate-controlled cab.

But the question remains how sustainable Singapore's network of more than 350 parks really is. The Gardens by the Bay, while iconic, is an artificial forest park—like a zoo for plants. Singapore has lost about half of all its animal species over the last two centuries. The same may happen to the rest of Southeast Asia by the end of this century.[36]

Much of Singapore's greenery cannot be eaten. The city-state produces less than 10 percent of its vegetables locally. Hence, it cannot rely on its own land to supply everyone with enough food. In an effort to rethink its food supply, it planned the Kranji Heritage Trail, which takes tourists on a path across the last remaining enclaves of local farming. However, skeptics consider urban land too expensive for agriculture.

Singapore is actively looking to solve this problem. Sky Greens, the world's first commercial vertical farm, consists of 120 aluminum A-frames, each with 38 tiers of planting and floating in a pond of water to capture rainwater and grow fish. Each A-frame revolves vegetables, like a Ferris wheel, so the tomatoes and other produce get bathed in sufficient water and light. The vertical farm produces half a ton of fresh vegetables per day.[37] With 6 million mouths to feed, the city will need to find space for several thousand more.

Singapore also faces an energy challenge to become truly sustainable. The Singapore economic miracle so far has worked because of air-conditioning, once singled out by Prime Minister Lee as one of two important factors to enable the city's success, with multicultural tolerance. "The first thing I did upon becoming prime minister was to install air conditioners in buildings where the civil service worked. This was key to public efficiency."[38] But this is a major energy guzzler, with 60 percent of all energy use of non-residential buildings used exclusively for air-conditioning.[39]

Still, Singapore has become a model for sustainable development worldwide. Cities wanting to adopt similar policies may find it difficult, though, since Singapore relies on strict enforcement. For instance, some elevators have been outfitted with urine detection devices to detect urine's scent, close the doors, and call the police. For its draconian penalties on offenses such as spitting and chewing gum, author William Gibson, no stranger to dystopian visions of the future, once described the city as "Disneyland with the Death penalty."

Nevertheless, as carbon emissions are heating up our atmosphere and urbanization is disrupting our natural system, Singapore remains a test bed for a more sustainable urban future—one centered around the greenest skyscrapers.

SINGAPORE PROVIDES AN EXAMPLE of how cities and skyscrapers can coexist with nature. Refusing to partake in a race to build the world's tallest tower, it has chosen to build the greenest instead. It has refined its vision from "Garden City" to "Biophilic City in a Garden," referencing biophilia, first termed by the eminent biologist E. O. Wilson in 1984. The biophilia hypothesis theorizes that humans are hardwired to connect with nature, even urbanites who may have never been out in a forest. This love of nature may literally be in our DNA—a result of millions of years of being attuned to forests and savannas, when our primate ancestors foraged for food and used trees to escape from predators.

A growing body of research supports the idea that exposure to green foliage increases attention spans and decreases stress levels. A study of 20,000 people found that those who spend at least 120 minutes in green spaces report substantially better health and psychological well-being.[40] It has even been shown that hospital patients with views of trees recover faster than those with views of brick walls.[41] People living closer to open green space are more effective in performing cognitive tasks or dealing with major life issues. This is why doctors in Japan prescribe "forest bathing."

With the majority of people living far from forests, urban planners and architects are increasingly considering how to bring "forests" into our cities and buildings. Even the City of London's bulwark of glass and metal towers is now home to grass and plants. On one of London's skyscrapers, a green wall rises 26 stories high. This building may not be as tall as the reflective Walkie-Talkie. But at least it's cooling down the nearby area instead of scorching it with reflected rays.

As an alternative to a race for the tallest vertical wall, a smarter target may be the building with the greenest plot ratio: the most landscape replaced in the sky from a piece of land. Singapore's Oasis Hotel holds the record for a hotel building, scoring more than 10, meaning the total area of the new plants replace ten times the site area. Meanwhile, the HDB is ramping up its efforts to integrate greenery into residences, forcing all new developments to get to a ratio of 4.5.

Already, the rise of the "tree scraper" is leading to unique innovations. The trouble with adding trees to skyscrapers is that they require balconies with lots of steel reinforcement, a carbon-intensive material. These green initiatives may outweigh the environmental cost to support them. To reduce this, trees may grow on walls and grow sideways—almost like a Magritte painting, a surrealist world turned on its side. GraviPlant is a tree that is placed horizontally in a rotating pot fixed onto the building's side. This constant rotation tricks the tree to grow sideways. This sideways tree is bushier than its vertical counterpart, since its constant rotations bathe it in more sunlight—not unlike a shawarma on a slowly turning spit. However, with each horizontally growing tree costing thousands of dollars, we are unlikely to see many sideway-growing trees on our buildings in the near future.

The most forward-thinking architects may want to collaborate not just with horticulturalists but with agriculturalists. Even our greenest cities waste energy on shipping food from thousands of miles away. We can minimize these "food miles" by incorporating agriculture into our buildings. This way, cities can be agriculturally productive instead of being consuming islands relying on a vast periphery of greenhouses and watersheds. Urban sprawl erases more farmland every year—we are losing an area as large as Denmark annually, with much of it in Asia.[42]

Already, restaurateurs and homeowners are pruning herbs, greens, and small fruits on the sides of buildings. Vertical gardens may even be coming to your high-rise apartment soon. Next time you need a clump of basil, you may be able to extend your hand from your apartment window, and pick from trellises or revolving trays. Or there could be entire skyscrapers dedicated to urban farming, so-called farmscrapers.

Other builders care less for the largest amount of vegetated areas than they do the number of species in their skyscrapers. The architect Jean Nouvel collaborated with a botanist to create One Central Park in Sydney. This glass skyscraper is checkered with patches of vegetation that house 250 native Australian flowers and plant species. Still, the tower to beat is arguably Milan's Bosco Verticale. In 2014, Italian architect Stefano Boeri designed two skyscrapers that contain not just 14,000 plants, but also 800 trees. Engineers needed to reinforce the balconies with steel to carry the extra weight, but the effort has paid off in unexpected ways. As a result of the towers' 100 tree and plant species, the project now creates an ecosystem of its own, attracting more than 20 different types of birds. Buildings like this may help undo some of the damage done by cities. However, critics point to the towers' disproportionate drain on the local water supply, all for a luxury building. As the French aristocracy once had a monopoly on green spaces, four centuries later there is still inequitable access to vegetation. We are waiting for eco-justice pioneers,

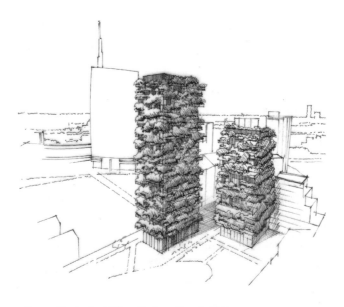

Bosco Verticale, Milan, Stefano Boeri, 2014

who will make clean air and urban park spaces a social issue, not just a green one.

In our age of biodiversity loss, the new challenge is to design buildings that will spawn their own ecosystems with the most species. Every living thing on this planet is a result of thousands if not millions of years of adaptation to different environmental conditions. Each can offer valuable insights for our own survival in the future. Biodiversity is our "library of life," a collection of 9 million "books" filled with knowledge. Now, as a result of urbanization and climate change, we are burning many of the "books" in our collective library.

Advocates around the world are making buildings more biodiversity-friendly, starting with roofs. "Brown" roofs, or biodiverse roofs, are designed to promote a natural habitat for birds, animals, plants, and invertebrates, without interference from people. To get to biodiverse buildings, architects will need to enter creative collaborations with botanists, biologists, zoologists, and marine scientists.

With all this flora and fauna vying for building walls, it may get crowded on skyscraper skins. Singapore is even considering whether it can use the sides of buildings for solar power. In their quest for more efficient solar panels, scientists are developing alternatives to the silicon wafers that make up conventional solar panels. Perovskite solar cells, made of man-made crystals called perovskites, can more efficiently convert light to electricity. Unlike traditional photovoltaics, perovskite solar cells can be sprayed or printed on wall surfaces. The Henn na Hotel in Nagasaki, Japan, already has a curved wall of perovskites that power the hotel's sign (the hotel is known for being ahead of the curve—it was once the world's first robot-staffed hotel, until the robotic concierges irritated customers). However, there may be a risk with the perovskites, since they contain small amounts of toxic lead. In case of solar panel failure, this poisonous material can wash into the soil and enter the food chain, causing major diseases. Once again, solving one problem leads to the next problem.

With buildings able to produce renewable energy, engineers must solve for another challenge. Buildings cannot share their potential energy surplus with the larger grid. Our existing power grids work in only one direction. They bring power from the plant to your outlet, not in reverse. The fundamentals of these grids date back to Thomas Edison's 1882 Pearl Street Station plant, an electricity-generating coal plant. As a result, buildings today cannot share resources—for instance, the way mycorrhizal fungal bridges share resources among trees.

With new technology, including sensors inside of smart meters, the regular one-way grid can become a two-way system. This smart grid can move the excess energy generated from the photovoltaics on your office's rooftop over to power another building.

There may be another lesson to learn from our ecosystem. There is no "waste" in nature, since one species' waste may be used productively by another species further down the food chain. Already, eco-industrial parks are creating "industrial ecologies," where companies use each other's waste products. In Kalundborg, Denmark, surplus heat from a power station is used to heat local homes and a fish farm. Sludge from the fish ponds gets an afterlife as fertilizer for neighboring farms. Gypsum, another by-product of the same power plant, is sent to wallboard manufacturers. These exchanges of "waste" increase the environmental efficiency of the whole.

Through resource sharing, so-called circular cities are increasingly eliminating waste in urban areas as well. For instance, Hammarby Sjöstad outside of Stockholm has an incinerator to convert combustible waste from households into electricity. In Barcelona, a system of pneumatic tubes runs below the streets to bring trash to an anaerobic digestion facility. These subways for trash more easily bring it to the facility. Then microorganisms break down the material into biogas, a renewable-energy source.

As cities are getting more circular, more sophisticated systems will need to manage and share resources. In 2014, Singapore launched a digital twin, a virtual version of the city called "E3A," "Everyone, Everything, Everywhere, All the Time." It displays 3D renderings of all the city's parks, buildings, and waterways, like the video game SimCity, based on real-time data such as energy use, pollution, and noise. The computer model can run virtual experiments and test policies before they are actually implemented. For instance, it can explore the impact of a new building or park on the shadows and wind flows. Systems like this may soon be able to calculate and evaluate the many opportunities for buildings to generate resources. This Big Brother may also be watching you, though, with computer vision monitoring your every garbage and sewage output.

With skyscrapers producing and sharing energy, food, species, and more, our world may look quite different. Imagine a day in Singapore a few years from now. You throw your recyclable trash into a chute, where a system of pneumatic tubes sucks it to the recycling plant, avoiding the need for a polluting garbage truck. As your trash arrives, robots scan and sort the material, separating paper from aluminum from plastics. Your waste is then turned into a resource, recycled as a new can or toothbrush. Your food waste is piped to an anaerobic digestion facility, where microorganisms break down the organic material, creating fertilizer and biogas that power the nearby vertical farm. With each storm, rainwater flows from the farm's roof through pipes and filters so it can be used to irrigate the crops. Self-driving freight delivery dollies roll the crops through underground tunnels to local stores, avoiding trucks polluting streets with noise and exhaust.

But it must be noted that as cities get smarter by collecting ever more data through more sensors, citizens' privacy will be at ever greater risk. Technological progress does not always lead to human progress; it can be quite the opposite. Utopia and dystopia are two sides of the same coin.

By 2050, over 68 percent of the world's population is expected to be living in cities. Designing green, circular, and zero-carbon buildings and cities has never been more important. Humans are the destroyers of the natural world, but we are also part of it. In order to sustain the planet that sustains us, city planners and architects need to lay an eco-friendly foundation. Creating innovative green spaces does not just benefit nature, it ensures our own survival.

The Future of Tall Buildings and a More Sustainable World

We are living in an extraordinary period of tall buildings. Never before have so many skyscrapers been constructed. A century ago, during the Roaring Twenties, a skyscraper boom changed the skyline of New York. Today, tall buildings find their way into cities across the globe, from London to Melbourne.

Not only are there more skyscrapers, now they are also getting increasingly taller. Supertall status—300 meters, about 1,000 feet— is no longer exceptional. Dozens of towers have already exceeded this bar, once set by the Eiffel Tower. In 1956, most people questioned the feasibility of Frank Lloyd Wright's proposal for a mile-high tower. Now today, if someone were to announce a "milescraper" to be built, maybe few people would be surprised? Perhaps the new bar will be the mile-high goal.

Skyscrapers have been around for more than a century. But they are no longer the same. They are not just about breaking new heights. They are getting built with new materials, like high-performance concrete and cross-laminated timber. They are standing more stable thanks to structural innovations, such as the buttressed core and tuned mass dampers. They are easier to navigate, thanks to faster, lighter, and smarter elevators. They are more efficiently cooled, supported by better mechanical systems and insulating envelopes. Their interiors are laid out differently, allowing for new ways for people to live, work, and socially interact with each other.

For environmental reasons alone, the way in which we design our tallest buildings has to change. Heat waves are increasing, droughts are extending, floods are occurring more frequently, and storms are getting more intense. We like to call these events natural disasters. Yet the vast majority of scientists agree they are actually human-induced, a result of climate change. Overwhelming evidence acknowledges the building sector's role in triggering climate change. Architects are bound by a kind of "Hippocratic Oath" for the public realm and environmental stewardship. Instead of just thinking of how to get taller-appearing structures, they should also aim to create buildings with smaller environmental footprints.

"It must be tall, every inch of it tall," Louis Sullivan, father of the skyscraper, wrote on the characteristic of the high office building in the late nineteenth century. "The force and power of altitude must be in it, the glory and pride of exaltation must be in it." For the person designing skyscrapers, he noted, "the problem of the tall office building is one of the most stupendous, one of the most magnificent opportunities that the Lord of Nature in his beneficence has ever offered to the proud spirit of man."[1] Sullivan aimed for tallness and for buildings designed to further emphasize their height, "without a single dissenting line," he wrote. The problem, for Sullivan, was how to impart graciousness to "this sterile pile, this crude, harsh, brutal agglomeration, this stark, staring exclamation of eternal strife."

Compared to a century ago, we now know how to build tall. Better materials, innovative structures, sophisticated air-conditioning, and faster elevators help us pierce the clouds. At the same time, mass transit, contextual design, microclimatic studies, and setback regulations can fit these towers into the city, without congesting streets and harming pedestrians. However, the overall question remains: why should we build tall today?

The environmental cost of tall buildings summons an existential challenge for skyscraper designers. The taller a building, the more energy it consumes. Compared to a low-rise building, a high-rise is hungrier for elevators, air-conditioning, electricity, and structure. With more material necessary to make tall structures withstand the blowing wind, the more energy is required to manufacture trusses, columns, and walls. As skyscrapers require extra structure, more mechanical equipment, larger pumps to move water up, and thicker elevator cores, the amount of usable floor space shrinks, creating even more waste. In other words, some towers are simply "too tall" for their own environmental good. Supertalls are like gas-guzzling Hummers on steroids.

Nevertheless, the race into the sky has propelled many innovations that we now benefit from. We discovered how to make stronger, lighter structures with less material. We learned how to make our buildings withstand the strongest winds with aerodynamic design and tuned mass dampers. We found how to move people up safely, quickly, and smoothly in elevators—and soon they will travel diagonally and sideways.

While skyscrapers over 1,000 feet tall are certainly not the world's most eco-friendly endeavors, humbler skyscrapers have significant environmental benefits. Stacking people in tall buildings brings them closer to more activities, reducing the per capita energy for transportation. With more people per unit of land, cities can invest more tax resources in infrastructure such as public transportation and high-quality open space. With people in compact multifamily buildings instead of single-family homes, there is less exterior surface area per apartment, with less heat or cooling lost to the outdoors. This helps reduce energy bills and carbon footprints.

Compare Manhattan with Phoenix. Manhattan has more skyscrapers, hence is more dependent on energy-intensive structures. Nevertheless, in the sprawling city of Phoenix, people need cars and miles and miles of energy and water lines. At the same time, dense urban development helps preserve land for environmental or agricultural benefits. Manhattan has more than twenty times the population density of Phoenix. If Manhattanites were spread out as much as people in Phoenix, they would require twenty times more land.

By 2050, 2.5 billion more urbanites are projected to be living in cities. In addition to new housing, this urban drive will require a massive increase of city infrastructure, including roads, sidewalks, parks, and transit, as well as city sewage, water, and power lines. This is like building a new New York City every single month for the next thirty years.[2] If we were to build Phoenixes, we would waste a lot more energy related to transportation and twenty times more environmental land. We don't need to tread so hard across the earth if we can rise into the sky.

Cities are heavy polluters, responsible for more than 70 percent of all carbon emissions.

Paradoxically, they are also the most sustainable thing we've got. While high-rise living may not be for everyone, many people actually prefer being in central cities, close to offices, restaurants, and transportation hubs. High-rises can be built in these locations. The more viable urban centers we have, the less our cities have to sprawl.

Tall buildings are an integral part of urban living for the future.

Sustainability and resilience are the challenges they will need to overcome. Global warming is raising sea levels. Meanwhile, the urbanization of coastal regions is putting more people in harm's way. Many of our city's buildings, skyscrapers included, are built too close to the shore. Piling people into tall structures increases societal risk, as we found out the hard way during recent flooding events and the COVID-19 pandemic. Sociologist Ulrich Beck claimed that modernization has created a "risk society" and now requires systematic efforts to handle uncertainty. Architects and urban planners can no longer design buildings and cities for a certain future. They will have to design with risk, probability, and resilience in mind. In fact, we can use these environmental challenges as an opportunity to create better buildings.

We need a new paradigm. We need to recalibrate the race for the tallest building. We should aim for buildings to be the greenest, with the most landscape, and the fewest carbon emissions. We need skyscrapers to generate the most renewable energy, produce the most food, and promote the healthiest environments for residents with the most biophilic benefits. We need tall buildings created with their entire life cycle in mind and to be disassembled in the end. We need the most resilient buildings able to withstand the ravages of climate change with not only practical design but also nods to aesthetics and beauty.

The good news is that many energy-efficient technologies and renewable-energy solutions are commercially available today, and they are already being applied. Renewable-energy supplies, cross-laminated timber, prefabricated modules, and super insulation are making a positive impact. Now we need a lot more. Despite advances in artificial intelligence and robotics, the built environment is still mostly a brick-and-mortar industry, capital-intensive and locally fragmented. It tends to lag behind. The vast majority of buildings will become greener only if regulations both force and incentivize them to do so.

Change is coming; however, there is potential for trouble. For instance, as much as outfitting buildings with sensors helps monitor and optimize energy efficiency, they are also adding prying mechanical eyes to our lives. Smart cities and buildings have the dangerous potential to invade privacy, dooming us to live in George Orwell's *1984*.

Experimentation comes with risk. Yet, without new proposals, we cannot change the status quo.

The story of the supertall tells us we should not let our surroundings be the sole influence on what we think is possible. As a species, we can push the laws of physics and create mind-bending structures. They are moon shots, and they may fail. If we succeed, we can create new wonders of the world.

If we can harness our ingenuity to build structures up in the clouds, then we can also create structures that are good for the planet down on earth.

ACKNOWLEDGMENTS

"A civilization is built by the successive contributions of genera-tions, leaning on each other like the stones of a building," French essayist André Frossard wrote. The same is true for *Supertall*.

I am grateful for the support of my literary agent Jud Laghi, who has been a thought partner from the beginning. I thank my editor Matt Weiland of W. W. Norton for his guidance, thoughtfulness, and enthusiasm. Huneeya Siddiqui, also at W. W. Norton, helped edit and steer the project to completion.

The striking ink illustrations are from the hands of David Dugas, whose skills with the French Curve made even the more curvilinear buildings come to life.

My friends reviewed various parts of the manuscript. I am grate-ful to Katie McElhenney for bringing her wits and creativity to this project. I thank Eva Warrick and Duane DeWitt for their construc-tive feedback. I also thank my fellow TED residents, in particular Ashwini Anburajan, Kat Mustatea, and Shani Jamila, who helped set the tone of the writing. My collaborations with Dan Kwartler of TED ED have inspired the story telling.

I am grateful to my loving wife, Rebecca, the cornerstone of our family. My late mother, a student of history, taught me to appreciate the lessons of the past. My late grandfather, a carpenter, put tools in my hands when I was a young child, allowing me to appreciate the craft of building early on.

This book is dedicated to Maxine, my daughter, as well as the new generation. May they elevate our society to new heights.

NOTES

INTRODUCTION: The Era of the Supertall

1. Blair Kamin, "Frank Lloyd Wright's Mile-High Skyscraper Never Built, But Never Forgotten," *Chicago Tribune*, May 28, 2018.
2. Jane King Hession and Debra Pickrel, *Frank Lloyd Wright in New York: The Plaza Years, 1954–1959* (Layton, UT: Gibbs Smith, 2007), 47.
3. Matthias Beckh, *Hyperbolic Structures: Shukhov's Lattice Towers—Forerunners of Modern Lightweight Construction* (Hoboken, NJ: John Wiley, 2015), 19.
4. Kevin O'Flynn, "A Cool Soviet-Era Landmark Faces Possible Destruction in Moscow," *BuzzFeed News*, March 31, 2014, accessed February 21, 2021, https://www.buzzfeednews.com.
5. United Nations Department of Economic and Social Affairs, "2018 Revision of World Urbanization Prospects," May 16, 2018, https://www.un.org.

ONE: The Building Block That Binds the World: Concrete

1. Christian Meyer, "The Greening of the Concrete Industry," *Cement and Concrete Composites* 31, no. 8 (2009): 601–5.
2. Morris Hicky Morgan and Herbert Langford Warren, *Vitruvius: The Ten Books on Architecture*, 1914, Book II, Chapter VI.
3. Ben Guarino, "Ancient Romans Made World's 'Most Durable' Concrete. We Might Use It to Stop Rising Seas," *Washington Post*, July 4, 2017.
4. Reyner Banham, *A Concrete Atlantis: US Industrial Building and European Modern Architecture, 1900–1925* (Cambridge, MA: MIT Press, 1986), 65.
5. Alfred O. Elzner, "The First Concrete Skyscraper," *Architectural Record* XV, no. 6 (1904): 515.
6. Henry Cowan, *The Master Builders: A History of Structural and Environmental Design from Ancient Egypt to the Nineteenth Century* (New York: John Wiley, 1977).
7. Kanita Dervic et al., "Suicide Rates in the National and Expatriate Population in

Dubai, United Arab Emirates," *International Journal of Social Psychiatry* 58, no. 6 (2012): 652–56.

8. James Alfred, "Burj Khalifa: A New High for High-Performance Concrete," *Proceedings of the Institution of Civil Engineers–Civil Engineering* 163, no. 2 (2010): 66–73.

9. Mir M. Ali, "Evolution of Concrete Skyscrapers: From Ingalls to Jin Mao," *Electronic Journal of Structural Engineering* 1, no. 1 (2001): 2–14.

10. Zoe Naylor, "Burj Team Keeps Floor-Cycle Promise," *Construction Week*, October 13, 2006.

11. Richard Merritt, "World's Tallest: The Creeping of the Burj Khalifa," *Duke, Pratt School of Engineering News*, January 12, 2020, accessed February 23, 2021, http://den.pratt.duke.edu/node/125.html.

12. Inken De Wit, "Understanding Corrosion Processes in Concrete," *Phys.org*, August 3, 2017, accessed February 23, 2021, https://phys.org/news/2017-08-corrosion-concrete.html.

13. Jing Ke et al., "Potential Energy Savings and CO_2 Emissions Reduction of China's Cement Industry," *Energy Policy* 45 (2012): 739–51.

14. Sophie Bushwick, "Bacteria-Filled Bricks Build Themselves," *Scientific American*, January 15, 2020.

15. Victor C. Li, "Bendable Concrete, with a Design Inspired by Seashells, Can Make US Infrastructure Safer and More Durable," *Michigan Engineering*, May 30, 2018, accessed February 23, 2021, https://news.engin.umich.edu/.

16. Francesca Pierobon et al., "Environmental Benefits of Using Hybrid CLT Structure in Midrise Non-Residential Construction: An LCA Based Comparative Case Study in the US Pacific Northwest," *Journal of Building Engineering* 26 (2019): 100862.

17. "Code Counts: New Wood-Concrete Hybrid Structural Floor System Tested for Mid- and High-Rise Tower Applications," *Think Wood*, March 26, 2018, accessed April 2020, https://www.thinkwood.com.

TWO: The Fight Against Sway: Wind

1. Harry G. Poulos, "Tall Building Foundations: Design Methods and Applications," *Innovative Infrastructure Solutions* 1, no. 1 (2016): 1–51.

2. Suger (Abbot of Saint Denis), and Gerda Panofsky-Soergel, *Abbot Suger on the Abbey Church of St.-Denis and Its Art Treasures* (Princeton, NJ: Princeton University Press, 1979), 101.

3. D. J. Stanley, "The Original Buttressing of Abbot Suger's Chevet at the Abbey of Saint-Denis," *Journal of the Society of Architectural Historians* 65, no. 3 (2006): 334–55.

4. Suger, *Abbot Suger on the Abbey Church of St.-Denis*, 22.

5. Vincent Curcio, *Chrysler: The Life and Times of an Automotive Genius* (Oxford: Oxford University Press, 2001), 426.

6. Yasmin Sabina Khan, *Engineering Architecture: The Vision of Fazlur R. Khan* (New York: W. W. Norton, 2004), 210.

7. Khan, *Engineering Architecture*, 119.

8. Blair Kamin, "Plain and Simple, Hancock Rules," *Chicago Tribune*, January 17, 1999.

9. Khan, *Engineering Architecture*, 230.

10. "Cloud-Piercer," *Ideas and Discoveries Magazine*, September 1, 2019, accessed February 24, 2021, https://www.pressreader.com/usa/id-magazine.

11. Clay Risen, "The Rise of the Supertalls: Engineering Advances Have Architects Striving for the Mile-High Skyscraper," *Popular Science*, February 15, 2013.
12. "A Tale of Two Towers," *Economist*, November 18, 2013.
13. Karissa Rosenfield, "In Progress: Shanghai Tower/Gensler," *Huffington Post*, May 4, 2012.
14. Kyoung Sun Moon, "Structural Design and Construction of Complex-Shaped Tall Buildings," *IACSIT International Journal of Engineering and Technology* 7, no. 1 (2015): 30–35.
15. Jon Galsworthy, "Rising to the Clouds with Confidence," *Structure Magazine*, June 2016.
16. Malcolm Millais, *Building Structures: Understanding the Basics* (New York: Routledge, 2017), 422.
17. Khan, *Engineering Architecture*, 361.
18. Sidewalk Labs, *Exploration 1: How to Design a Timber Building That Can Reach 35 Stories or More*, January 24, 2020, accessed April 17, 2020, https://sidewalklabs.com.
19. Nicholas A. Yaraghi et al., "A Sinusoidally Architected Helicoidal Biocomposite." *Advanced Materials* 28, no. 32 (2016): 6835–44.

THREE: The Race to the Top: Elevators

1. Kheir Al-Kodmany, "Tall Buildings and Elevators: A Review of Recent Technological Advances," *Buildings* 5, no. 3 (2015): 1070–104.
2. Statista, 2021, "Number of New Elevator and Escalator Installations in China from 2001 to 2019," March 17. Retrieved from https://www-statista-com.ezproxy.cul.columbia.edu/statistics/1202047/new-installations-elevators-escalators-china/.
3. Lyu Qiuping, Li Bin, and Guan Guifeng, "China Focus: Elevators Boost Life in Old Communities," *Xinhua*, June 9, 2020.
4. CTBUH, *2019 Tall Building Year in Review*, 2020, accessed April 17, 2020, https://www.skyscrapercenter.com/.
5. Spencer Klaw, "'All Safe, Gentlemen, All Safe!' The Ups and Downs of the Invention That Forever Altered the American Skyline," *American Heritage* 29, no. 5 (1978).
6. Rem Koolhaas, *Delirious New York: A Retroactive Manifesto for Manhattan* (New York: Monacelli Press, 2014).
7. Marcus Vitruvius Pollio, *The Ten Books on Architecture*, trans. Morris Hicky Morgan (Adamant Media Corporation, 1960).
8. Christopher Klein, "1,500 Years Later, Killer Animal Elevator Returns to Colosseum," *History*, September 1, 2018, accessed April 18, 2020, https://www.history.com.
9. "Steam Boilers Are Exploding Everywhere," *Scientific American*, March 1881.
10. Lee Edward Gray, *From Ascending Rooms to Express Elevators: A History of the Passenger Elevator in the 19th Century* (Elevator World Inc., 2002), 239.
11. Jeannot Simmen and Uwe Drepper, *Der Fahrstuhl: Die Geschichte Der Vertikalen Eroberung* (Prestel, 1984), 55.
12. Gray, *From Ascending Rooms to Express Elevators*, 258.
13. Andreas Bernard, *Lifted: A Cultural History of the Elevator* (New York: New York University Press, 2014), 252.
14. Bernard, *Lifted*, 240.
15. Nick Paumgarten, "Up and Then Down," *The New Yorker*, July 28, 2014.

16. Antonio Sant'Elia, "Manifesto of Futurist Architecture," *Futurist Manifestos* (London: Thames & Hudson, 2009).
17. Stephen Graham, "Super-Tall and Ultra-Deep: The Cultural Politics of the Elevator," *Theory, Culture & Society* 31, no. 7-8 (2014): 239–65.
18. Angus K. Gillespie, *Twin Towers: The Life of New York City's World Trade Center* (New Brunswick, NJ: Rutgers University Press, 1999), 84.
19. Research In China, *Global and China Elevator Industry Report, 2019–2025*, November 2019.
20. Adam Taylor, "The Surprisingly Cutthroat Race to Build the World's Fastest Elevator," *Washington Post*, January 4, 2017.
21. "Super Technology for Ultra-High-Speed 20.5 m/s," Mitsubishi Electric Global website, accessed April 18, 2020, https://www.mitsubishielectric.com/elevator/innovations/worlds_fastest.html.
22. Michael McCann and Norman Zaleski, *Deaths and Injuries Involving Elevators and Escalators* (Center to Protect Workers' Rights, 2006).
23. Taylor, "Surprisingly Cutthroat Race."
24. David Malone, "Shanghai Tower Nabs Three World Records for Its Elevators," *Building Design and Construction*, February 8, 2017, accessed April 18, 2020, https://www.bdcnetwork.com/.
25. Brad Nemeth, "Energy-Efficient Elevator Machines," *ThyssenKrupp Elevator AMS Energy Monitoring Program*, 2011.
26. Roald Dahl, *Charlie and the Chocolate Factory* (London: Penguin Books, 1964).
27. Peter Kenter, "Revolutionary MULTI Elevator System Headed for Canada," *Daily Commercial News*, September 12, 2018, accessed February 24, 2021, https://canada.constructconnect.com/.

FOUR: The Cooling Effect: Air-Conditioning

1. C. L. Barnhardt, *The American College Dictionary* (New York: Random House, 1957), 1431.
2. Thibaut Abergel et al., *Towards a Zero-Emission, Efficient, and Resilient Buildings and Construction Sector: Global Status Report* (Paris: UN Environment Programme and the International Energy Agency, 2018).
3. US Energy Information Administration, "Annual Energy Outlook 2020," Washington, DC (2020).
4. Charles W. Carey Jr., *American Inventors, Entrepreneurs, and Business Visionaries* (Infobase Publishing, 2014), 59.
5. Margaret Ingels and Willis H. Carrier, *Willis Haviland Carrier, Father of Air Conditioning* (Country Life Press, 1952), 18.
6. Ingels and Carrier, *Willis Haviland Carrier*, 20–21.
7. Willis H. Carrier, "Rational Psychrometric Formulae. Carrier," *Journal of the American Society of Mechanical Engineers* 33, no. 11 (1911): 1311–50.
8. Reyner Banham, *Architecture of the Well-Tempered Environment* (Chicago: University of Chicago Press, 1984), 171–72.
9. "From Benjamin Franklin to John Lining, 17 June 1758," *Founders Online*, National Archives, accessed February 23, 2021, https://founders.archives.gov/documents/Franklin/01-08-02-0023.
10. "Refrigeration for Towns and Cities by Street Mains," *Scientific American*, September 5, 1891.

11. Ingels and Carrier, *Willis Haviland Carrier*, 88.
12. Stan Cox, *Losing Our Cool: Uncomfortable Truths About Our Air-Conditioned World (and Finding New Ways to Get Through the Summer* (New York: The New Press, 2010), 9.
13. Marsha Ackerman, *Cool Comfort: America's Romance with Air-Conditioning* (Washington, DC: Smithsonian Institution Press, 2010), 121.
14. Banham, *Architecture of the Well-Tempered Environment*, 11.
15. Daniel A. Barber, "Heating the Bauhaus: Understanding the History of Architecture in the Context of Energy Policy and Energy Transition" (Philadelphia: Kleinman Center for Energy Policy, 2019), 5.
16. Pam Belluck, "Chilly at Work? Office Formula Was Devised for Men," *New York Times*, August 4, 2015.
17. Tom Wolfe, *From Bauhaus to Our House* (New York: Macmillan, 1981), 2.
18. Patrick Sisson, "How Air Conditioning Shaped Modern Architecture—And Changed Our Climate," *Curbed*, May 9, 2017, accessed February 24, 2021, https://archive.curbed.com.
19. Christopher S. Wren, "International Symbol of Neglect; U. N. Building, Unimproved in 50 Years, Shows Its Age," *New York Times*, October 24, 1999.
20. "The Tallest Structure: Burj Khalifa," *33 Extremes on Earth* (Authors of Wikipedia, 2013).
21. Clay Risen, "How to Heat and Cool a Supertall," *Architect Magazine*, December 1, 2015, accessed April 19, 2020, https://www.architectmagazine.com.
22. "The Challenge: Tall and Super-Tall Buildings: HVAC," *Consulting Specifying Engineer Magazine*, July 28, 2014, accessed April 19, 2020, https://www.csemag.com.
23. Carly Fordred, "A Tall Order," *HVAC&R Nation*, March 2010.
24. John Robert, "City Planning in Frankfurt, Germany, 1925–1932: A Study in Practical Utopianism," *Journal of Urban History*, November 1, 1977.
25. Donald Aitken, "The 'Solar Hemicycle' Revisited: It's Still Showing the Way," *Wisconsin Academy Review* 39, no. 1 (1992–93): 33–37.
26. Buckminster Fuller, "The Case for a Domed City," *St Louis Post-Dispatch*, September 26, 1965.
27. "Steve Rose. "'His Inner Circle Knew about the Abuse': Daniela Soleri on Her Architect Father Paolo," *The Guardian*, February 29, 2020.
28. Saskatchewan Research Council, "A Closer Look at the Saskatchewan Conversation House and Four Others," April 17, 2018. Accessed October 3, 2021, https://www.src.sk.ca/.
29. Sophia Epstein, "Everyone Needs to Stop Building Giant Glass Skyscrapers Right Now," *Wired*, November 11, 2019, accessed February 24, 2021, https://www.wired.co.uk.
30. American Birds Conservatory, *Glass Collisions: Preventing Bird Window Strikes*, accessed February 24, 2021, https://abcbirds.org/glass-collisions/.
31. Kamaljit Singh et al., "The Architectural Design of Smart Ventilation and Drainage Systems in Termite Nests," *Science Advances* 5, no. 3 (2019).

FIVE: The Rules That Shape Skylines: London

1. Jill Jonnes, *Eiffel's Tower: The Thrilling Story Behind Paris's Beloved Monument and the Extraordinary World's Fair That Introduced It* (London: Penguin Books, 2009).
2. Alexandra Lange, "Seven Leading Architects Defend the World's Most Hated Buildings," *New York Times*, June 5, 2015.

3. Markus Moos, "From Gentrification to Youthification? The Increasing Importance of Young Age in Delineating High-Density Living," *Urban Studies* 53, no. 14 (2016): 2903–20.

4. Paul Swinney and Andrew Carter, "London Population: Why So Many People Leave the UK's Capital," *BBC News*, March 18, 2019, https://www.bbc.com/.

5. Goldman Sachs Investment Research, "Millennials Coming of Age," 2015, accessed February 24, 2021, https://www.goldmansachs.com/insights/archive/millennials/.

6. Hannah Arendt, *The Human Condition* (Chicago: University of Chicago Press, 1998), 26–27.

7. Nadine Moeller, *The Archaeology of Urbanism in Ancient Egypt: From the Predynastic Period to the End of the Middle Kingdom* (Cambridge: Cambridge University Press, 2016).

8. Aristotle, *Politics*, Book 2, Section 1267b, trans. H. Rackham, *Aristotle in 23 Volumes* (Cambridge, MA: Harvard University Press, 1944).

9. Morris Hicky Morgan and Herbert Langford Warren, *Vitruvius: The Ten Books on Architecture*, 1914, Book I, Chapter VI.

10. Derek Keene, "Towns and the Growth of Trade," *The New Cambridge Medieval History*, eds. David Luscombe and Jonathan Riley-Smith, 4: 76.

11. Spiro Kostof, *The City Shaped: Urban Patterns and Meanings Through History* (London: Thames & Hudson, 1999).

12. Robert Fishman, *Bourgeois Utopias: The Rise and Fall of Suburbia* (New York: Basic Books, 1987), 10.

13. Friedrich Engels, *The Condition of the Working Class in England* (Oxford: Oxford University Press, 1993), 65.

14. James Thomson, "The City of Dreadful Night," *National Reformer*, March 22, 1874.

15. Norma Evenson, *Paris: a Century of Change*, 1878–1978. (New Haven, CT: Yale University Press, 1979): 149.

16. Michael Jameson, "Domestic Space in the Greek City State," in *Domestic Architecture and the Use of Space: An Interdisciplinary Cross-Cultural Study*, ed. Susan Kent (Cambridge: Cambridge University Press, 1990), 103.

17. Francesca Street, "Are London's Protected View Corridors Still Relevant Today?" CNN, September 21, 2017.

18. Historic England, *London's Image and Identity—Revisiting London's Cherished Views*, 2018, 9. https://historicengland.org.uk.

19. P. J. Waldram and J. M. Waldram, "Window Design and the Measurement and Predetermination of Daylight Illumination," *The Illuminating Engineer* 90 (1923): 96.

20. *Hkruk II (CHC) Limited v Marcus Alexander Heaney*, England and Wales High Court 2245 (Ch), Leeds, 2010.

21. Reuters, "Travel Picks: 10 Top Ugly Buildings and Monuments," November 14, 2008, accessed February 26, 2021, https://www.reuters.com.

22. James Holloway, "The Shard's Bleeding Edge: Anatomy of a 21st Century Skyscraper," *Ars Technica*, December 4, 2011, accessed February 24, 2021, https://arstechnica.com.

23. Greater London Authority, *The London Plan 2004: Spatial Development Strategy for Greater London*, February 2004, 5.

24. Rowan Moore, "The Shard: A Symbol of Towering Ambition," *The Guardian*, January 29, 2011.

25. John L. Gray, "Report to the First Secretary of State: Land Adjoining London

Bridge Station, at St Thomas Street and Joiner Street, London SE1," July 2003, APP/ A5840/V/02/1095887.

26. Reuters, "Global Daily Forex Trading at Record $6.6 Trillion as London Extends Lead," September 16, 2019.

27. Alain de Botton, "London Is Becoming a Bad Version of Dubai," video, *The Guardian*, July 14, 2015, accessed February 21, 2021, https://www.theguardian.com/.

28. Emma Glanfield, "London's Walkie Talkie Skyscraper Is Voted the Ugliest and Most Hated Building in Britain as It Is Branded 'A Gratuitous Glass Gargoyle Graffitied on the Skyline,'" *Daily Mail*, September 2, 2015.

29. Martin Spring, "Prince Charles Attacks Rash of New 'Carbuncles,'" *Building*, February 1, 2008, accessed February 26, 2021, https://www.building.co.uk/.

30. David Thame, "Manhattan on the Irwell: Skyscraper Cluster Is Really Happening," *Bisnow*, January 21, 2020, accessed February 21, 2021, https://www.bisnow.com/.

31. Oliver Wainwright, "Welcome to Manc-hattan: How the City Sold Its Soul for Luxury Skyscrapers," *The Guardian*, October 21, 2019.

32. Rowan Moore, "The Shard: A Symbol of Towering Ambition," *The Guardian*, January 29, 2011.

33. Guy de Maupassant, *The Complete Works of Guy de Maupassant: The Wandering Life and Short Stories*, trans. Alfred de Sumichrast (New York: Pearson Publishing Co., 1910), 1.

34. New London Architecture, *London Tall Buildings Survey 2020*, April 2020.

35. Westminster City Council, *Building Height Study*, June 2019, accessed February 24, 2021, https://www.westminster.gov.uk/.

36. Guy Debord, *The Society of the Spectacle* (Good Press, 2021).

37. Simon Usborne, "'The Building Creaks and Sways': Life in a Skyscraper," *The Guardian*, February 4, 2017.

38. Ben Waber, Jennifer Magnolfi, and Greg Lindsay, "Workspaces That Move People," *Harvard Business Review* 92, no. 10 (2014): 68–77.

39. European Union and UN Habitat, "The State of European Cities 2016: Cities Leading the Way to a Better Future" (2016).

SIX: The Competition for Air Rights: New York

1. David Albouy, Gabriel Ehrlich, and Minchul Shin, "Metropolitan Land Values," *Review of Economics and Statistics* 100, no. 3 (2018): 454–66.

2. Hendrik Hartog, *Public Property and Private Power: The Corporation of the City of New York in American Law, 1730–1870* (Ithaca, NY: Cornell University Press, 1989), 163.

3. *Parker v. Foote*, 19 Wend. 309 (N.Y. Sup. Ct. 1838).

4. Thomas Jefferson, *Notes on the State of Virginia*, ed. Thomas P. Abernethy (New York: Harper Torchbook, 1964), 158.

5. John M. Levy, *Contemporary Urban Planning* (New York: Routledge, 2016), 9.

6. "The Crust at Chicago," *New York Times*, October 18, 1891, 4.

7. Peter Hall, *Cities of Tomorrow: An Intellectual History of Urban Planning and Design Since 1880* (Hoboken, NJ: John Wiley, 2014), 35.

8. Hall, *Cities of Tomorrow*, 37.

9. Henry James, *The American Scene* (London: Chapman & Hall, 1907), 80.

10. H. A. Caparn, "The Riddle of the Tall Building: Has the Skyscraper a Place in American Architecture?" *The Craftsman* 10 (1906): 476–88.

11. Louis Jay Horowitz, *The Towers of New York: The Memoirs of a Master Builder* (New York: Simon & Schuster, 1937).

12. William Atkinson, *The Orientation of Buildings or Planning for Sunlight* (New York: John Wiley, 1912).

13. Carol Willis, *Form Follows Finance: Skyscrapers and Skylines in New York and Chicago* (New York: Princeton Architectural Press, 1995).

14. George S. Chappell, "The Sky Line," *The New Yorker*, July 12, 1930, 67.

15. Jason Barr, "Skyscraper Height," *Journal of Real Estate Finance and Economics* 45 (2012): 723–53.

16. Rem Koolhaas, *Delirious New York: A Retroactive Manifesto for Manhattan* (New York: Monacelli Press, 2014), 155.

17. Daniel Okrent, *Great Fortune: The Epic of Rockefeller Center* (London: Penguin Books, 2003), 399.

18. Department of City Planning, New York City, "Transcript of Public Hearing Before the City Planning Commission, March 14 & 15, 1960, in the Matter of Comprehensive Amendment of the Zoning Resolution of the City of New York," New York, 1960.

19. Alden Whitman, "Mumford Finds City Strangled by Excess of Cars and People," *New York Times*, March 22, 1967.

20. William H. Whyte, *City: Rediscovering the Center* (New York: Doubleday, 1988), 233.

21. William H. Whyte, *The Social Life of Small Urban Spaces* (New York: Project for Public Spaces, 1980), 15.

22. "Conveyance and Taxation of Air Rights," *Columbia Law Review* 64, no. 2 (1964): 338–54.

23. William B. Harvey, "Landowners' Rights in the Air Age: The Airport Dilemma," *Michigan Law Review* 56 (1957): 1313.

24. Donald J. Trump and Tony Schwartz, *Trump: The Art of the Deal* (New York: Ballantine Books, 2009), 164.

25. Steven Johnson, "Blueprint for a Better City," *Wired*, December 1, 2001.

26. Robin Finn, "The Great Air Race," *New York Times*, February 22, 2013.

27. Kerry Burke, Greg B. Smith, and Corky Siemaszko, "Crane Collapse in Midtown Manhattan as Hurricane Sandy Storms into the East Coast," *New York Daily News*, October 29, 2012.

28. Michael Kimmelman, "Seeing a Need for Oversight of New York's Lordly Towers," *New York Times*, December 22, 2013.

29. "City Council Considers Task Force to Study Impact of Megatowers Casting Shadows on Central Park," *CBS New York*, November 12, 2015, accessed February 27, 2021, https://newyork.cbslocal.com/.

30. Quoctrung Bui and Jeremy White, "Mapping the Shadows of New York City: Every Building, Every Block," *New York Times*, December 21, 2016.

31. Justin Davidson, "Giants in Our Midst," *New York Magazine*, September 12, 2013.

32. "Ultra High Net Worth Individual (UHNWI)," *Investopedia*, October 26, 2017, accessed May 16, 2020.

33. Council on Tall Buildings and Urban Habitats, "CTBUH Year in Review: Tall Trends of 2020," *CTBUH Journal*, no. 1 (2021).

34. Kristina Shevory, "Cities See Another Side to Old Tracks," *New York Times*, August 2, 2011.

35. Kevin Loughran, "Parks for Profit: The High Line, Growth Machines, and the Uneven Development of Urban Public Spaces," *City & Community* 13, no. 1 (2014): 49–68.

36. Michael Kimmelman, "Hudson Yards Is Manhattan's Biggest, Newest, Slickest

Gated Community. Is This the Neighborhood New York Deserves?" *New York Times*, March 14, 2019.

37. Oliver Wainwright, "Horror on the Hudson: New York's $25bn Architectural Fiasco," *The Guardian*, April 9, 2019.

38. Azi Paybarah, "They Were Going to Build a Wall at Hudson Yards. Then Came the Backlash," *New York Times*, January 15, 2020.

39. Nikolai Fedank, "Viñoly's Jetsons-Esque Skyscraper at 249 East 62nd Street Revealed, Upper East Side," *New York YIMBY*, February 20, 2018, accessed February 27, 2021, https://newyorkyimby.com/.

40. The Municipal Arts Society of New York, *The Accidental Skyline*, October 2017.

41. David B. Caruso, "One World Trade Center Named Tallest U.S. Building," Associated Press, November 12, 2013.

42. Stefanos Chen, "The Downside to Life in a Supertall Tower: Leaks, Creaks, Breaks," *New York Times*, February 3, 2021.

43. Henry Petroski, "Super-Tall and Super-Slender Structures: Skyscrapers with Smaller Footprints Require Countermeasures to Wind and Sway," *American Scientist* 107 (2019): 342–45.

44. Andrew Lawrence, "The Skyscraper Index: Faulty Towers," Property Report, Dresdner Kleinwort Benson Research (January 15, 1999).

45. Emily Badger, "Density Is Normally Good for Us. That Will Be True after Coronavirus, Too," *New York Times*, March 24, 2020.

46. Lewis Mumford, *The City in History: Its Origins, Its Transformations, and Its Prospects* (Boston: Houghton Mifflin Harcourt, 1961), 34.

47. Alan Berube, *MetroNation: How US Metropolitan Areas Fuel American Prosperity*, Metropolitan Policy Program at Brookings, 2007.

48. Enrico Berkes and Ruben Gaetani, "The Geography of Unconventional Innovation," *Rotman School of Management Working Paper* 3423143 (2019).

49. Luís Bettencourt, José Lobo, Dirk Helbing, Christian Kühnert, and Geoffrey West, "Growth, Innovation, Scaling, and the Pace of Life in Cities," *Proceedings of the National Academy of Sciences* 104, no. 17 (2007): 7301–6.

SEVEN: The Transit System That Supports Skyscrapers: Hong Kong

1. Martin V. Melosi, "The Automobile Shapes the City," *Automobile in American Life and Society* (University of Michigan, 2010), http://www.autolife.umd.umich.edu/Environment/E_Casestudy/E_casestudy2.htm.

2. Eran Ben-Joseph, *ReThinking a Lot: The Design and Culture of Parking* (Cambridge, MA: MIT Press, 2012).

3. David Biello, "No Such Thing as a Free Parking Spot," *Scientific American*, January 9, 2011.

4. John Tibbetts, "Coastal Cities: Living on the Edge," *Environmental Health Perspectives* 110, no. 11 (2002): A674–A681.

5. Le Corbusier, *The City of To-morrow and Its Planning*, trans. Frederick Etchells (New York: Dover, 1987), 3.

6. "Navvies: Workers Who Built the Railways," *National Railway Museum*, May 16, 2018, https://www.railwaymuseum.org.uk/.

7. Le Corbusier, *The City of To-morrow and Its Planning*, 131.

8. Michael Southworth and Eran Ben-Joseph, *Streets and the Shaping of Towns and Cities* (Washington, DC: Island Press, 2013).

9. "A Service of World-Class Quality," *MTR*, https://www.mtr.com.hk.

10. Peter Newman and Jeff Kenworthy, "Peak Car Use: Understanding the Demise of Automobile Dependence," *World Transport Policy and Practice* 17, no. 2 (2011): 35–36; UN Habitat, Planning and Design for Sustainable Urban Mobility: Global Report on Human Settlements 2013 (Nairobi: UN Habitat, 2013).

11. International Energy Agency, "CO_2 Emissions from Fuel Combustion by Sector in 2014, in CO_2 Emissions from Fuel Combustion," *CO_2 Highlights 2016*, Paris, 2016.

12. Lawrence D. Frank, Martin A. Andresen, and Thomas L. Schmid, "Obesity Relationships with Community Design, Physical Activity, and Time Spent in Cars," *American Journal of Preventive Medicine* 27, no. 2 (2004): 87–96.

13. World Health Organization, "Global Status Report on Road Safety," Geneva, 2009.

14. Robert Putnam, "Bowling Alone: America's Declining Social Capital," *Journal of Democracy* 6, no. 1 (1995): 65–78.

15. Jan Gehl, *Life Between Buildings: Using Public Space* (Washington, DC: Island Press, 2011).

16. Chris Rissel et al., "Physical Activity Associated with Public Transport Use—A Review and Modelling of Potential Benefits," *International Journal of Environmental Research and Public Health* 9, no. 7 (2012): 2454–78.

17. Emporis, "Most Skyscrapers," accessed February 27, 2021, http://www.emporis.com/statistics/most-kyscrapers.

18. John Calimente, "Rail Integrated Communities in Tokyo," *Journal of Transport and Land Use* 5, no. 1 (2012): 19–32.

19. Environment Bureau, Hong Kong, "Energy Saving Plan for Hong Kong's Built Environment 2015~2025+," Hong Kong, 2015.

20. Hong Kong Transport Department, "Summary of Key Statistics," accessed October 5, 2021, https://www.td.gov.hk/en/road_safety/road_traffic_accident_statistics/2019/.

21. Stefan Al, "Hong Kong's Transit-Oriented Podium-Tower Development," in *Vertical Urbanism*, ed. Zhongjie Lin (London: Routledge, 2018), 33.

22. Robert C. Schmitt, "Implications of Density in Hong Kong," *Journal of the American Institute of Planners* 29, no. 3 (1963): 210–17.

23. "Gloves to Muffle Stadium Applause," *South China Morning Post*, October 13, 1994.

24. Jeffrey Hardwick, *Mall Maker: Victor Gruen, Architect of an American Dream* (Philadelphia: University of Pennsylvania Press, 2010), 216.

25. Yuko Okayasu, "Hong Kong Tops Global Ranking of Most Expensive Shopping Streets," *Cushman & Wakefield*, November 14, 2019, accessed February 21, 2021, https://www.cushmanwakefield.com/.

26. Stefan Al, "Mall City: Hong Kong's Dreamworlds of Consumption," in *Mall City* (Honolulu: University of Hawaii Press, 2020), 1–20.

27. Nikolaus Lang et al., "Can Self-Driving Cars Stop the Urban Mobility Meltdown?" Boston Consulting Group, July 8, 2020, https://www.bcg.com.

28. Ben Hamilton-Baillie, "Towards Shared Space," *Urban Design International* 13, no. 2 (2008): 130–38.

29. SpaceX, "Hyperloop Alpha," August 12, 2013, accessed February 27, 2021, https://www.spacex.com.

EIGHT: The Greening of Vertical Cities: Singapore

1. Emporis, "Skyline Ranking," accessed February 26, 2021, https://www.emporis.com/.

2. William H. Stiebing and Susan N. Helft, *Ancient Near Eastern History and Culture* (New York: Routledge, 2017).

3. Wasana De Silva, "Urban Agriculture and Sacred Landscape: Anuradhapura Sacred City, Sri Lanka," *WIT Transactions on Ecology and the Environment* 217 (2018): 941–52.

4. Michal Artzy and Daniel Hillel, "A Defense of the Theory of Progressive Soil Salinization in Ancient Southern Mesopotamia," *Geoarchaeology* 3, no. 3 (1988): 235–38.

5. De Silva, "Urban Agriculture and Sacred Landscape," 950.

6. Robin Wall Kimmerer, *Braiding Sweetgrass: Indigenous Wisdom, Scientific Knowledge and the Teachings of Plants* (Minneapolis: Milkweed Editions, 2015), 229.

7. Cecil Konijnendijk, *The Forest and the City: The Cultural Landscape of Urban Woodland* (Berlin: Springer, 2008).

8. George V. Profous, "Trees and Urban Forestry in Beijing, China," *Journal of Arboriculture* 18, no. 3 (1992): 145–54.

9. H. W. Lawrence, *City Trees: A Historical Geography from the Renaissance Through the Nineteenth Century* (Charlottesville: University of Virginia Press, 2008), 24.

10. H. W. Lawrence, *City Trees*, 26.

11. "Keur van Amsterdam," Amsterdam, 1454.

12. Lucie Laurian, "Planning for Street Trees and Human–Nature Relations: Lessons from 600 Years of Street Tree Planting in Paris," *Journal of Planning History* 18, no. 4 (2019): 282–310.

13. "Pruning the Trees on the Promenades of Paris," *Scientific American Supplement* 1167 (May 14, 1898).

14. Judson Kratzer, "Savannah, Georgia: The Lasting Legacy of Colonial City Planning. Teaching with Historic Places," National Park Service, 2002, accessed February 27, 2021, https://www.nps.gov/.

15. Kevin J. Hayes, *The Road to Monticello: The Life and Mind of Thomas Jefferson* (Oxford: Oxford University Press, 2012), 467.

16. John M. Golby and A. William Purdue, *The Civilisation of the Crowd: Popular Culture in England, 1750–1900* (London: Batsford Academic and Educational, 1984), 102.

17. Frederick Law Olmsted, "The Yosemite Valley and the Mariposa Big Tree: A Preliminary Report," *Landscape Architecture* 43, no. 1 (October 1952): 12–25.

18. Stephen Smith, "On Excessive Death Rates Among Children Under Five Years of Age, and on Measures of Prevention," *The Sanitarian* 3, no. 31 (1875): 289.

19. John Charles Olmsted, Olmsted Brothers Landscape Architects, and Frederick Law Olmsted, "Portland: Report of the Park Board: Portland, Oregon (1903)," 1903.

20. The City of Portland, "Forest Park," accessed February 27, 2021, https://www.portland.gov/parks/forest-park.

21. "Stream Restoration Will Cool Down Seoul," *The Dong-A Ilbo*, August 12, 2005, accessed February 27, 2021, https://www.donga.com/.

22. Nate Berg, "Green Infrastructure Could Save Cities Billions," *City Lab*, April 24, 2012, accessed February 27, 2021, https://www.bloomberg.com/.

23. City of Seattle, "Ordinance 122311," *Seattle City Council Bills and Ordinances*, December 21, 2006, accessed February 27, 2021, http://clerk.seattle.gov/search/ordinances/122311.

24. Claudia Copeland and Nicole T. Carter, "Energy-Water Nexus: The Water Sector's Energy Use" (2014).

25. Robert Herman, "Green Roofs in Germany: Yesterday, Today and Tomorrow," *Green Roof Conference Proceedings*, Chicago, 2003.

26. Lee Kuan Yew, *From Third World to First: 1965–2000: The Singapore Story* (New York: HarperCollins, 2000), 176.

27. "'Garden City' Vision Is Introduced," *History SG*, March 2015, accessed February 27, 2021, https://eresources.nlb.gov.sg/.

28. Lee, *From Third World to First*, 178.

29. Lee, *From Third World to First*, 177.

30. Lee, *From Third World to First*, 663.

31. Lee, *From Third World to First*, 179.

32. Lee, *From Third World to First*, 98.

33. Amy Kolczak, "This City Aims to Be the World's Greenest," *National Geographic*, February 28, 2017, accessed February 27, 2021, https://www.nationalgeographic.com/.

34. Urban Redevelopment Authority, "Greenery," accessed February 27, 2021, https://www.ura.gov.sg/.

35. Richard T. Corlett, "The Ecological Transformation of Singapore, 1819–1990." *Journal of Biogeography* (1992): 411–20.

36. Navjot S. Sodhi et al., "Southeast Asian Biodiversity: An Impending Disaster," *Trends in Ecology & Evolution* 19, no. 12 (2004): 654–60.

37. Kheir Al-Kodmany, "The Vertical Farm: A Review of Developments and Implications for the Vertical City." *Buildings* 8, no. 2 (2018): 24.

38. Lee Kuan Yew, "The East Asian Way—With Air Conditioning," *New Perspectives Quarterly* 26, no. 4 (2009): 111–20.

39. Kian Jon Chua, Siaw Kiang Chou, W. M. Yang, and Jinyue Yan, "Achieving Better Energy-Efficient Air Conditioning—a Review of Technologies and Strategies," *Applied Energy* 104 (2013): 87–104.

40. Mathew P. White et al., "Spending at Least 120 Minutes a Week in Nature Is Associated with Good Health and Wellbeing," *Scientific Reports* 9, no. 1 (2019): 1–11.

41. Roger S. Ulrich, "View Through a Window May Influence Recovery from Surgery," *Science* 224, no. 4647 (1984): 420–21.

42. Eric F. Lambin and Patrick Meyfroidt, "Global Land Use Change, Economic Globalization, and the Looming Land Scarcity," *Proceedings of the National Academy of Sciences* 108, no. 9 (2011): 3465–72.

CONCLUSION: The Future of Tall Buildings and a More Sustainable World

1. Louis H. Sullivan, "The Tall Office Building Artistically Considered," *Lippincott's Magazine* (March 1896): 406.

2. Architecture 2030, "Why the Building Sector?" accessed February 27, 2021, https://architecture2030.org/.

INDEX

Page numbers in *italics* refer to illustrations.